Listening to Landscape

Listening to Landscape

Hauntology and the Echoes of Albion

Phil Hubbard

BLOOMSBURY ACADEMIC
NEW YORK • LONDON • OXFORD • NEW DELHI • SYDNEY

BLOOMSBURY ACADEMIC
Bloomsbury Publishing Inc, 1359 Broadway, New York, NY 10018, USA
Bloomsbury Publishing Plc, 50 Bedford Square, London, WC1B 3DP, UK
Bloomsbury Publishing Ireland, 29 Earlsfort Terrace, Dublin 2, D02 AY28, Ireland

BLOOMSBURY, BLOOMSBURY ACADEMIC and the Diana logo are trademarks of Bloomsbury Publishing Plc

First published in the United States of America 2026

Cover design by Louise Dugdale
Cover image © Lucy Hubbard

Library of Congress Cataloging-in-Publication Data

Names: Hubbard, Phil author
Title: Listening to landscape : hauntology and the echoes of Albion / Phil Hubbard.
Description: [1.]. | New York: Bloomsbury Academic, 2025. |
Includes bibliographical references and index. | Summary: "Moving deftly between cultural theory, musicology and geography, Listening to Landscape serves as a primer on hauntronica that defines the genre as uniquely concerned with questions of landscape and Englishness. However, it argues that 21st-century hauntronica is not mere nostalgia, but should be considered a provocation that helps us imagine a new radical politics for the Anthropocene. In making this argument, it speaks to urgent questions of national identities in the post-Brexit era, offering a distinctive take on the way contemporary culture deals with the ghosts and memories of Albion"-- Provided by publisher.
Identifiers: LCCN 2025017992 | ISBN 9798765112922 hardback |
ISBN 9798765112960 paperback | ISBN 9798765112946 pdf | ISBN 9798765112939 epub
Subjects: LCSH: Music--Environmental aspects--England | Music and geography--England | Soundscapes (Music)--England--History and criticism | Cultural geography--England | Landscapes--England
Classification: LCC ML3917.G7 H83 2025 | DDC 306.4/8420942--dc23/eng/20250715
LC record available at https://lccn.loc.gov/2025017992

ISBN: HB: 979-8-7651-1292-2
ePDF: 979-8-7651-1294-6
eBook: 979-8-7651-1293-9

Typeset by Deanta Global Publishing Services, Chennai, India

For product safety related questions contact productsafety@bloomsbury.com.

To find out more about our authors and books visit www.bloomsbury.com and sign up for our newsletters.

Contents

Figures

Acknowledgements

This book has involved a journey into music and landscapes often unfamiliar to me. For that reason, I must thank those who initiated me and guided me through, especially the many musicians, sound artists, producers and performers who gave up their time to talk to me and granted me permission to include their images in this volume. In no particular order, these included Will Yates, Tommy Perman, Gordon Chapman-Fox, Craven Faults, Richard Hanscomb, Rebecca Denniff, Matt Saunders, Jim Ghedi, Martyn Stonehouse, David Chatton-Barker, Ben Powell, Alan Outram, Alex Cargill, Paul Broome, Luke J Murray, Mat Handley, Matt Shaw, Realf Heygate, Liz Hanks, Ryan Hooper, David Jaycock, Craig Whitby, Michael Tanner, Chris Sharp, Mark Peters, Nick Sales, Jonathan Sharp, Greg Wye, Ben Holton, Colin Morrison, Mark Jones, Frances Castle, Mark S. Williamson, Eric Heath, Dave Clarkson, Matt Shenton, Oliver Cherer, Robin Saville, Matt Davies, Stephen Cracknell, Simon Tong, Hayley Suviste, Benjamin Harrison, Richard Adams, Joey and Kate from White Flowers, Jez Butler, Jim Jupp, Hannah Peel, Dominic O'Donoghue and George Rayner-Law from Degradation, Suzy Mangion, Michael Schofield and, especially, Justin Hopper. I have not been able to draw on all the interviews conducted with these artists, writers and performers, but their input was always gratefully received. For their ideas and inspired writing, thanks also to Ben Platt, Julian Holloway and Jonathon Turnbull. Ultimately, I also wish to thank the colleagues at King's College for allowing me to present some of this work to them, as well Eleanor, Lucy and Livvy for being so patient with me when deadlines loomed. I am also grateful to Bloomsbury Music for their prompt and enthusiastic response to the original book proposal, especially Leah Babb-Rosenfeld. Finally, this book is dedicated to Tim Edensor, whose infectious enthusiasm for interrogating the landscape should be an inspiration to us all.

Author's note

Readers are encouraged to enjoy *Listening to Landscape* while listening to this mix of the music discussed in the book: https://www.mixcloud.com/gypjh2000/listening-to-landscape-mix/

Many of the independently produced tracks discussed in this book can also be purchased or streamed via Bandcamp, with an indicative playlist at: https://bndcmpr.co/c938735a

Foreword

The road on which I live, in a beautiful, uncanny village in exurban East Anglia, runs in a dead-straight line for exactly 2 kilometres before curving, just slightly, around a rare incline. While busy in the morning and afternoon rush hour, it is largely unused from suppertime onwards, and often when sleepless I will lie still and listen for the nocturnal punctuations created by a lone, speeding car at 3.00 am. Speed demons are emboldened by the endless visibility of our road, a kind of rural *Serene Velocity*, in which the 'dub' repetitiveness of the fields and houses creates a hallway-tunnel in which time and place blur and speed becomes unknowable.

That 3.00 am unknowable speed is the sound of Brexit Britain, a dull propulsion through the revered countryside in darkness, without a thought or a glance, just forward motion through the past. Equally, it is the sound of enclosure – the movement two centuries ago in which England's commonly held lands were parcelled into lots and suddenly 'owned' by a landlord. Farming and controlling these lands then required the consolidation of myriad trackways and footpaths into single straight roads like the one outside my house: 'enclosure roads', as my late elderly neighbour still called them, even in the 2010s.

Brexit and enclosure haunt the English countryside, and that sound I hear at night is a kind of ghost story, a story of the unspeakable changes wrought upon the countryside, told by the sound of the land itself. *Listening to the Landscape* might also be seen as a book of ghost stories, woven by Phil Hubbard into a pantheon of myths: a kind of grimoire for understanding contemporary England through the undead and uncanny landscapes created through music. Ghost stories are, of course, the *lingua franca* of the English landscape, being the only stories that can capture the elusive quality of 'Englishness', which the landscape always points towards but never reaches. To be English means to believe in and talk of ghosts – whether those exist as a tale of a beast scouring the moorland countryside, or as a terrace song about a long-departed footballer whose presence might fire one's team into action.

As Phil Hubbard alludes to, Brexit, too, is already a ghost story: it is the tale of Merlin's Return, of a once-great kingdom given new life – being 'unleashed' – and becoming respected and feared throughout the world. We cannot understand

Brexit without understanding the oak tree that grows on Merlin's grave any more than we can understand Britain without knowing its landscapes and their multiplicity of meanings. Landscape is a tool that the inhabitants of England use to negotiate their brittle relationship with time. Through the tool of landscape, they understand place not simply as what is there – a forest or a river or a dry cleaner or a housing estate – but as an environment that includes change over time, and the ghosts of those ill-described lives that were once there (or which we might imagine were there, in the case of recent arrivals like myself).

In *Listening to Landscape* Phil Hubbard describes one of the ways we tell our English ghost stories today: through hauntological, and time-warped sound. We who operate within these circles – and I refer to the author here, as well as the incredible cast of characters, musicians and artists within these pages, and, as a reader, to you as well – are involved in an often-quiet search for the soul of this nation. We're answering the time-honoured question, what does it mean to live in an old country? Does it mean we are swimming in ancient waters and that's where it ends? Does it mean dreams of modernity, chased by frumpy classicism? Does it perhaps mean pagan ceremonies at the bonfire on the heath? Or does it mean discovering for ourselves what kinds of confusions and unanswerables – undecidables – might exist within the human, non-human and more-than-human world we inhabit?

Paul Connerton (2009: 122), in *How Modernity Forgets*, points out that 'Compared with all previous history, the life expectancy of people and that of buildings is now reversed'. For 'buildings', feel free to read not just the urban but all landscapes. What once seemed permanent is today merely an ephemeral tale. The result is that an 'old country' like England has become a land of ruins and future-ruins, whether that means Stonehenge or the London Gherkin or my long straight road that stands in for the pathways it subsumed. And ruins are twin symbols: they stand for something that once was but no longer is. But this also means that the way things are is not necessarily the way they will always be.

In this book, Phil Hubbard searches for ruins in the way we explore place through music and, when he finds them, reads them in a manner that reminds us: we still have choices. To misquote the filmmaker Nicholas Ray, we look for the landscapes we might believe in, and when we cannot find them, we create them for ourselves.

Justin P. Hopper, February 2025

1

Soundtracking the Nation

Since the close-run 2016 Brexit referendum, where the populace voted 52:48 to leave the EU, the UK has seemed more divided than united. The vote revealed significant splits between young and old – with youthful, more educated voters appearing more likely to vote Remain than Leave – but it also exposed important geopolitical fractures.[1] Scotland and Northern Ireland, for example, resoundingly voted Remain, with Brexit delivered, demographically and culturally, by a core of English voters who appeared disenfranchised by conventional politics. In the wake of the largely unexpected outcome, political commentators and journalists hence descended on strongly Brexit-supporting towns in England – Mansfield, Boston, Stoke – in their attempt to understand why the Leave campaign's mantra of 'taking back control' had such an appeal in these traditionally working-class communities. Based on their observations, they theorized a relationship between deprivation, de-industrialization and political disenfranchisement, the term 'left behind places' quickly becoming part of the political argot, informing the gestural politics of the Levelling Up agenda intended to promote the development of the North and Midlands.[2]

This has led to the idea that the UK is splintering, riven by regional divisions of North versus South, town versus country, struggling periphery versus wealthy metropolitan core.[3] The post-COVID-19 recession, and the disastrous 2022 Tory budget, appears to have entrenched these divides, creating a 'cost of living' crisis which the Keir Starmer-led Labour government seems at a loss to solve. Fuel poverty has risen for the first time in decades, and more people are reliant on food banks than ever before (including some in professions like teaching and nursing). England has some of the least affordable housing in the world, the most expensive public transport in Europe, and many high streets have more than a quarter of their shops lying vacant. The NHS is chaotic, with a massive waiting list growing longer by the day. The nation's environmental goals, including reducing food waste, improving water consumption and reversing species decline, have been consistently missed. Sewage is routinely pumped into rivers, and flood defences struggle to cope with 'extreme' weather that's becoming more regular.

Given this context of *meta-crisis*, and the widespread perception that English MPs in Westminster 'broke' Britain, it's unsurprising questions of national identity have become fraught. Devolutionary pressures are now evident not just in the struggles for Northern Irish, Scottish and Welsh independence but also in campaigns aiming to increase the powers given to the mayoral combined authorities, which now cover 40 per cent of England's population. This push to independence became even stronger after the Brexit vote itself, with the gap between Westminster and political leaders in the North widening during the shambles of the COVID-19 lockdown when boozy Conservative party staff partied in Whitehall while the nation observed strict quarantine.

Combined, the fallout of Brexit, COVID-19 and the political crisis in Westminster has contributed to much of the population of England distancing itself from identifying as English. Notably, between the Census of Population of 2011 and 2021, the number referring to themselves as English declined dramatically from 58 per cent to 15 per cent of the population, many evidently preferring to refer to themselves as British rather than English. In contrast, Scottishness seems a more inclusive identity, with almost all ethnic minority populations in Scotland more likely to claim a Scottish identity than English ethnic minorities do in England. For many, Englishness has become a dirty word, a disreputable nationalism equated with intolerance and isolationism: among the 15 per cent who still proclaim English identity there is unquestionably a sizeable racist group who feel the country has been overrun by migrants and believe the refugees arriving at the south coast in boats are 'illegals'. Here, the mantra of 'take back control' associated with the Leave campaign appears to have encouraged such exclusionary beliefs.[4] With Nigel Farage becoming an MP for the right-wing populist Reform Party in 2024, and the Conservative opposition lurching further to the right under leader Kemi Badenoch, this rhetoric is not disappearing: the anti-immigrant 'riots' of July 2024 that followed the fatal stabbing of three young girls in Southport were arguably given legitimacy by those right-wing politicians who, following the lead of former Home Secretary Suella Braverman, described the arrival of refugees at the South coast as an 'invasion' and vowed to 'push back the boats'.[5] Significantly, in the wake of the victories of Farage's Reform Party at local elections in 2025, Keir Starmer's Labour Party decided the time was right to impose even stronger migration rules to prevent the nation becoming a 'land of strangers' and suggested all migrants need to learn to speak English.

Exclusionary nationalism seems to have eclipsed multicultural cosmopolitanism across the political spectrum, with those arguing against it dismissed as part of the 'tofu-eating wokerati'.

This febrile context makes any examination of the representation of the English landscape a fraught undertaking. Yet it is also necessary. Nation-building is, after all, a socio-cultural project as much as a political achievement, with images of the national landscape having been especially prominent in the discourses of those arguing for Leave. For example, while the preservation of England's 'green and pleasant land' was often implicit in Boris Johnson's Churchillian allusions to the 'broad sunny uplands' that awaited after the divorce from Europe, sometimes this invocation was more explicit. EU policies, we were told, had encouraged destructive farming practices, resulting in the collapse of native bird and insect numbers, with the Common Agricultural Policy poorly suited to the specificities of English countryside. Leave campaigners posited English nature as vulnerable to 'foreign' influences of various kinds: urban encroachment, climate change, predatory invasive species and so on. Here, alarmist stories linking migration to ecological decline were often used by the right-wing press to whip up public concern about the porosity of the national border: rather than framing concern about refugee arrivals in explicitly racist terms, these articulated anxieties about immigration in environmental terms. For example, the Leader of the House of Commons, Leave campaigner Chris Grayling, suggested there were simply too many people in Britain, with the English countryside being 'ruined by migrant homes'.[6]

Questioning the rural idyll

In the dark politics of Brexit, the idea of England's lost rural idyll shone bright, illuminating the rhetoric of those arguing for a divorce from Europe. Leave campaigners tapped into a hazy imagery of warm beer, cricket pitches and rolling, green fields to construct the idea of a timeless island nation needing to be freed from overseas influence. They understood it's easy to provoke outrage when the English countryside appears threatened by things deemed to belong elsewhere, whether this is the bureaucracy and red tape of the EU Common Agricultural Policy, the extension of the Trans-European Transport Network through swathes of Middle England via High Speed Two, or the presence of invasive 'European super fleas', stink bugs, 'Euromoths' and giant continental dormice in the English countryside.[7] Their innately conservative 'ecological Englishness' was emotionally resonant for many living in the rural, despite the fact that the nation-state is an inadequate scale-framing for tackling the most pressing issues of our time, not least global climate change.[8]

Yet the idea that the countryside encompasses all that's good about England, and needs protection from 'foreign influence', has been challenged. For example, there are now multiple commentators suggesting the iconographic landscapes of Englishness are not innately English, but the product of complicated histories of migration and diaspora.[9] Anita Sethi's (2021) *I Belong Here: Walking the Backbone to Britain*, for example, reclaims the English countryside from white domination, arguing that for many non-white citizens, the idealized rural remains frustratingly out of reach.[10] Noreen Masud's (2023) *A Flat Place* has been acclaimed for recounting experiences of racial trauma in the fens of Cambridgeshire, a place she argues has much in common with rural Pakistan despite the seeming opposition of English rurality and Asian Otherness.[11] John Akomfrah's three-screen video installation *Vertigo Sea* (2015) uses a collage of film, history and literary references to explore the long-standing connections between racial exploitation and landscape of the Tamar Valley in Cornwall, highlighting the limitations of versions of Englishness that exclude racialized others.[12] For geographer Divya Tolia-Kelly, such post-colonial engagements with rural landscapes usefully unravel 'narratives of "strangers", "others" and "blackness"' subsumed in dominant representations of rurality and tradition.[13]

At the same time, the shelves of the average bookshop positively groan under the weight of the 'new nature writing' exploring the complexity of English nature. Following in the wake of Richard Mabey's exploration of the 'unofficial' countryside, this includes poetry, eco-literature and photography focusing on the less celebrated landscapes of England: industrial ruins, wastelands, the coastal margins.[14] While much of this is not explicitly about race or national identity, such engagements draw on theories of hybridity to stress English landscapes always have a diverse ecological and cultural history. Registering these varied histories displaces nativist nationalism's claim on the English landscape, exposing it as a racialized and classed fantasy.

Yet despite the obvious mismatch between images of a rural idyll and the multicultural realities of modern Britain, idealized rural landscapes are still recurrently deployed by politicians attempting to manufacture consensus about what it means to live on an 'island nation'.[15] Here, landscape becomes a synecdoche for the nation, a powerful *visual* referent serving as an important locus of attachment and sentiment. For such reasons, landscape art has been instrumental in shaping ideas of nationhood. Constable's and Gainsborough's paintings of rolling green countryside, for example, have long provided the quintessential vision of Englishness, with the seeming stability and continuity of the rural ways of life they depict instilling a pride in the pastoral picturesque, and a rejection of the complex present in favour of a comforting and bucolic antiquity. As Peter Taylor argues, the celebrated landscapes of rurality in art effectively fused 'an actual (or aspired) sovereignty, the history of a territory, as well as a selection of routinized habits, events, memories, narratives and iconographies related to the purported national identity'.[16] This occurred because the visual consumption of landscape in art – as well as related traditions of photography and film – is intensely pleasurable, provoking sometimes pronounced feelings of belonging, contentment and even pride.

But where does music fit into all of this? One answer is that landscape is more than a 'way of seeing', with the meaning of the English landscape constructed through *embodied encounters* as well as through its visual representation. John Wylie, for example, argues that rather than just studying images of landscape, we also need to examine the diverse materialities and sensibilities through which it's experienced. As he explains, landscape is something not just seen, but also *heard*.[17] Likewise, David Matless argues 'landscape is worked through all senses, though with attention to the ways in which hierarchies of sense operate; the truths of sight, stories overheard, sounds and textures'.[18] Listening to the landscape hence provides an 'additional channel of knowledge, producing insights into scale, materiality and landscape morphology that are not available through other ways of knowing'.[19]

Landscape, music and nationalism

What, then, is to be gained by listening to landscape? One simple response is that if we listen carefully, we can discern the processes involved in the making of our world: the crashing of pebbles on a shore, for example, or the gentle groan of an icy lake as it expands and contracts. We hear the animals with whom we share our world: birds that chirp, insects that buzz, cattle that low. If we listen particularly attentively, we might also begin to realize that the ambient soundscape has its own rhythms. These are sometimes discordant or jarring, but on occasion they create a *eurhythmia* as the sounds of human and non-human activity entwine to create distinctive and pleasing soundscapes. The English coast, for example, is readily invoked by the sound of waves regularly lapping on the beach, children playing in the surf, squawking seagulls wheeling in the sky above and the occasional melodic tinkle of an ice cream van in the distance.

In this sense, the representation of landscape in painting and photography is usefully complemented by its sonification, with the distinctive qualities of sound playing a very particular role in summoning the qualities of the landscape. Here, field recording represents an important and expanding tradition, involving 'the production, circulation, and playback of audio recordings of the myriad soundings of the world: the sounds of animals, birds, cities, machines, forests, rivers, glaciers, public spaces, electricity, social institutions, architecture, weather – anything and everything that vibrates'.[20] Field recording encourages us to listen closely, acknowledging the co-constitutive role of human and non-human agents in environmental processes. It can also be a source of inspiration for musical composition, with an increasing number of artists incorporating field recording and sampled sounds in electronic music-making. Simultaneously, new technologies are enabling more performers to take their practices into the landscape itself, recording in situ and allowing the atmosphere of place to bleed into the compositional process. As Michael Gallagher argues, this opens space for 'the play of imagination, fantasy and reverie' and 'amplifies the partiality and provisionality of landscapes as sites that are always contested, always unfinished'.[21]

The distinction between field recording, sound art and music is becoming arguably more blurred because of this. But even when music does not include any element of field recording or situated performance, it can still invoke the scale, beauty, texture and mood of landscape at a distance.[22] This is because music can trigger feelings and emotions via the remembrance of places we have been, or experienced: music is, after all, the soundtrack to our lives.[23] Evidentially, the extent to which music evokes a particular place and time depends on our positionality, memory and biography. Not all music is heard or reacted to in the same way. But there is clearly some music which has, through shared performance and broadcast, become indelibly associated with certain places: for example, Seattle grunge, K-Pop or Merseybeat. Particular genres, styles or types of music evoke towns or regions, becoming a synecdoche that stands for the whole. It's for this reason that particular composers, styles and pieces of music have become enrolled in ideological projects of nation-making, contributing to imaginings of collective identity: witness the singing of the English national anthem at sporting events or the televised spectacle of the Last Night of the Proms, where classical music performs a 'banal nationalism' connecting citizens to their country.[24]

The English pastoral tradition has been particularly important here, emerging in the first few decades of the twentieth century as a powerful musical evocation of the 'green and pleasant land'. Reaching its apogee in the immediate wake of the First World War, Gustav Holst, Frederick Delius and Ralph Vaughan Williams' pastoralism drew on traditions of English folk and choral music, rejecting continental European influences in favour of music that acted as a salve for a nation reeling from the repercussions of the Great War.[25] Famously, Vaughan Williams claimed that music, 'above all other arts', was the ultimate expression of the 'soul of the nation'. His 'Lark Ascending' – perhaps the quintessential encapsulation of English pastoralism – underlined this by topping Classic FM's nation's favourite classical music poll for eleven consecutive years at the beginning of the twenty-first century. As Richard King notes in his study of the links between music and nationhood, Vaughan Williams was no naturalist, but his evocation of a skylark soaring over green fields of England alluded to the seemingly immutable relation between nature and Englishness.[26] In so doing, it contributed to a classical pastoral tradition that mapped English national identity onto an idealized rural landscape, one where villages nestled in rolling downlands, and farm labourers lived in harmony with their masters.

Pastoral compositions were generally uncomplicated, slow, and gently lyrical: flutes, pipes and woodwind evoked a rural way of life, echoing the poetry of Donne and Milton and the art of Gainsborough and Constable. This had tremendous popular appeal in the inter-war years. For those returning from the horrors of the First World War, the evocation of Arcadia was particularly alluring, taking them away from the memory of the Flanders landscape torn apart by machines of destruction. But in a wider sense, pastoralism's romantic idealization of the countryside served as a counterpoint to a European Modernity often depicted as profoundly alien to English traditions: it emphasized the tranquillity of rural landscapes that appeared threatened by the rush to industrial capitalism and the machine age. As several commentators have noted, there were clear links between its popularity and the rise of organizations fighting for the protection of rural England.[27]

Beyond the pastoral?

Vaughan Williams, Delius and Holst were composers who drew on Tudor courtly and folk idioms, rejecting the continental European influences dramatically transforming classical music in the early twentieth century in favour of an Englishness that symbolized security and solidity. As George Revill stresses, pastoralism was not a musical style per se, but a 'powerful set of musical resources' which allowed composers and performers to map imaginaries of nationalism onto a series of mythologized places: orchards, gardens and pastures in particular.[28] It turned its back on the metropolis, depicting old England as a demi-paradise, unblemished by Modernity and infused with the 'Grace of God', an idea that gained popularity thanks to the ruralism espoused in the 1930s by H. J. Massingham, a Cotswolds-based writer firmly opposed to any notion of 'progress'.[29] In this sense, pastoral music worked alongside the rural writing, art and poetry of the period to construct a notion of British identity that marginalized Northern Ireland, Scotland and Wales in favour of a 'deep' Englishness imaginatively located in lowland, Southern England.

This implies that pastoral music was both religiously and politically conservative. Indeed, Vaughan Williams' biographer Eric Saylor notes that many in the early twentieth century musical *avant garde* took issue with the aesthetic and cultural values pastoralism promoted, particularly its continued celebration of folk music and its desire to develop a robust culture of 'national music' based on rurality. Pastoralism was often labelled escapist, sentimental and mawkish: Elisabeth Lutyens famously dismissed it as 'cow-pat music'.[30]

But such associations are not immutable. Despite its enduring popularity, not all those hearing 'Lark Ascending' today will recognize the version of Englishness it connotes, in the same way that not all those participating in the Last Night of the Proms uncritically embrace its patriotism (notably, in 2023 the BBC fielded complaints that many in the Albert Hall crowd were sporting EU flags as a comment on Brexit). Indeed, there are numerous examples of contemporary musicians adopting the tropes of pastoralism to subvert and undermine the myth of the English rural idyll. P. J. Harvey, for example, won the 2011 Mercury Music Prize for *Let England Shake*, an album showing the idealization of the English countryside obscures a long-standing history of slavery and bloodshed.[31] Brought up in rural West Dorset on the family farm, Harvey often works local folklore and dialect into her music, having referenced the landscape of the county on albums including *White Chalk* (2007), *Let England Shake* (2011) and *I Inside the Old Year Dying* (2023). Abigail Gardner argues Harvey's work is pastoral not in the sense it depicts the English countryside as an idyllic place of retreat and return but because it uses that landscape as a setting for stories of 'disappointment, death and war'.[32] *White Chalk*, for instance, borrowed many of the idioms of folk music and reworked them into stories of isolation and frustration ('White chalk hills are all I've known / White chalk hills will rot my bones'), while *Let England Shake* suggested the soil, chalk and earth of England is stained with the blood of multiple conflicts, a visceral recounting of the enduring emotional power of the rural.[33] *I Inside the Old Year Dying* is based on a seasonal poem cycle (*Orlam*) composed in local dialect, suggesting a gothic pastoral in which sexuality and mortality entwine ('Know you every tear in these woods / Know you every place of good and not-good'). Its mixture of feverish folk, warped electronica and treated field recordings again suggests little continuity with the musical traditions of classical pastoralism, and often her voice is distorted beyond recognition, but the album's sensual evocation of the rural suggests the bittersweet melancholy of fading youth and the proximity of 'life and death', themes always present in the pastoral tradition.[34]

Other musicians have pushed this deconstruction of pastoralism even further. Gazelle Twin – the alter ego of Canterbury-born Elizabeth Bernholz – splices the traditional pastoral influences of folk and early music (recorders, tambourine and harpsichord) with the acid house that reverberated through the English countryside in the 1990s. On *Pastoral* (2018), she mocks those romanticizing rural England, depicting a countryside home to rural thugs, joyriders and blue-rinsed UKIP-supporters, its dystopian rurality of suspicion and anxiety informed by Bernholz's relocation from Brighton to rural Leicestershire in the immediate aftermath of Brexit:

> I hadn't set out to write an album on the themes of rural life at all. The birth of my son slammed us into a strange and alien new role – with heightened anxiety, and a very protective, slightly paranoid and fearful state (not helped by post-natal depression), which seemed to jar horrendously with the very idyllic and quiet setting we were living in . . . The birdsong, the parks and friendly well-behaved neighbours are often a pleasant thing, but most of the time as a new mum, I felt like I was bristling with dread. I suddenly felt very aware of the violence and injustice that lay behind almost everything I deemed *nice* . . . Casual aggression, casual racism, casual xenophobia, casual evil, all readily in the air once again.[35]

Appearing in videos as a jester-imp resplendent in Adidas sportswear in the colours of the cross of St. George (Figure 1.1), Bernholz's *Pastoral* project saw her caricature recognizable English rural types – provincial curtain-twitchers, NIMBYs, racists – with a typical lyric summarizing their values: 'Serve yourself, take pride / Work hard, trust no one / Stay in alert all the time'. Gazelle Twin's music is uncompromising and sometimes scary avant-pop that encapsulates rurality in the twenty-first century, not portraying a timeless countryside but one where contemporary prejudices are built on those of the past ('Over The Hills', e.g. reworks the song 'Over the Hills and Far Away', from George Farquhar's 1706 play *The Recruiting Officer*, about soldiers being press-ganged into fighting in the Wars of the Spanish Succession).

The collaboration between multi-instrumentalist/composer Hannah Peel and poet Will Burns – *Chalk Hill Blue* (2019) – likewise dwells on some of the dysfunction of the contemporary rural (Figure 1.2). The album is named

Figure 1.1. Gazelle Twin's *Pastoral.* Artwork by Gazelle Twin, Typography/Graphics by Jonathan Barnbrook.

Figure 1.2. Hannah Peel and Will Burns. Courtesy: Chris Turner.

after an endangered English butterfly found in relative abundance near Burns' home in Wendover, Buckinghamshire, an 'unspoilt' English market town at the foot of the Chilterns on the ancient Ridgeway between Avebury and Tring. On the album, Burns' flatly delivered narrative suggests a complicated sense of place in which birdsong, the babble of chalk streams and the 'vernacular of flint and brick' is the backdrop to pub car park arguments, drinking sessions and fracturing relationships. *Chalk Hill Blue* recounts slack afternoons, lost keys, plastic bags of dog shit, fishing trips, hazy mornings of frying eggs and all the time intimates at personal and collective loss. The death of a relative in a nursing home is hinted at ('Out of Doors', 'Swallowing'), while the unseasonal cold weather leaves a pile of feathers and a bird's corpse on the canal towpath ('Change'). Both the cycle of life and the rhythms of nature are evoked by tracks referencing the seasonal year: summer is 'mosquito time', hay is cropped in the fading day and birdsong is 'everywhere' in the gloom of a spring dawn. A crust of dead ladybirds congeals on the kitchen windowsill during a heatwave

('Summer Blues'). The synth arpeggios that bookend many of the tracks seem to represent the cycles of nature repeating, but also a life going nowhere, the protagonist both lost and found in Middle England.

In the later novel *The Paper Lantern* (2021) – named for his parent's pub where he spent the COVID-19 lockdown holed up, divorced and aged nearly forty – Burns attests to the closed horizons of rural life, but also describes the redemptive power of the locality:

> I was up at four with the sun and the birdsong. My days had taken on some of the rhythm of these non-human things, beginning in the dim first light . . . Early mornings on top of the hill, the birdsong rich and varied in the beech woods, the bluebells shimmering, the wild garlic pungent in the cool air . . . Half a dozen red kites glided slowly past me . . . Their forms in the moving air were hypnotic, something between chaos and supreme control. Economy of movement, absolute discretion in each flurry of breeze in their flight feathers. It was beyond worship, what I felt watching those birds. This was the keen desire for the world, felt anew.[36]

In *Chalk Hill Blue* there are moments evoking this transcendence, with Peel's electronic violin and piano producing elegiac pastoral passages interspersed between the electronic hum and throb: performed live at the Barbican in London, Peel and Burns were joined by a woodwind section and sound artist Chris Watson, who augmented the album with field recordings of chalk streams and insect choruses.

Yet in *Chalk Hill Blue*, the natural world appears threatened by human settlement, property and development, not least the High Speed Two (HS2) rail link between London and the Midlands laying waste to the hills and woods so dear to Burns. Tellingly, Wendover is adjacent to the Chequers estate, the Prime Minister's country home, which remains untouched by the construction work, but the train line nonetheless goes through the heartlands of mid-Buckinghamshire. Construction of the 405-metre Wendover viaduct has destroyed ancient woodlands, many cottages have been demolished to make way for 'cut and cover' tunnels and watercourses have been diverted by the construction works, threatening flooding and loss of groundwater. The village's cricket pitch, where Burns recalls fielding on the boundary while watching mistle thrushes grubbing around the sight screens, has been destroyed and the promised replacement long delayed by planning issues.

Chalk Hill Blue is also a comment on rural gentrification. As Burns describes, much of rural Buckinghamshire is 'close enough to London for the high rollers of the city to escape to', leading to local pubs being full of 'a dull and rapacious middle-class who propagate themselves on a particular kind of mutual congratulation and self-satisfaction'.[37] He continues by noting 'there are people around my village with proper old Buckinghamshire accents (like my Grandad), but there's also lots of London overspill and that kind of placeless Southern Englandness to it as well'.[38] Here, Burns captures the essence of a form of twenty-first-century gentrification that places value and status on living close to nature but which often destroys that nature through forms of over-consumption, the privatization of the commons and accumulation through dispossession.[39]

In many ways, *Chalk Hill Blue* exposes contemporary rural geographies of extraction, efficiency and profit and the emergence of a countryside moulded by what Burns describes as the 'small-minded reactionary, nationalistic nervous breakdown' of Brexit.[40] As Burns notes, England is a country in love with its rural past, a dream kept alive in the present despite bearing little relation to the realities of contemporary village life. Indeed, the idea that rural living involves some sort of idyllic communion with nature appears increasingly out of kilter with the post-productivist countryside. As Burns argues, 'You only have to see the disjunct between how "rural" the country as a whole sees itself and the actual state of our environment with its totally degraded river system, a depleted insect and bird population, and uplands denuded of natural features'.[41] In the sleeve notes for *Chalk Hill Blue*, Burns suggests there are simply 'not enough butterflies, not enough lizards or water voles or fish, certainly not enough birds' anymore. Yet he suggests there is much to be learnt about alternative rural futures by engaging with the remaining vestiges of the countryside: 'It's broken in many ways and . . . we're causing the changes. But what's sad about the degradation of our times is that we can still see the potential nature of real places when we come up against them. These old paths, these old stories, these old buildings. We don't need them for nostalgia or for some artificial sentimental reverie, we need them to function as engines for our own epochal story-making'.[42]

Musically ambitious and socially astute, *Chalk Hill Blue* suggests pastoralism remains relevant to the landscapes of the twenty-first century but needs to be twisted into new forms to acknowledge the contemporary realities of English society. Alongside the work of P. J. Harvey and Gazelle Twin, the music of Hannah Peel and Will Burns builds on longer-standing histories of pastoralism as a musical idiom that represents Englishness, subverting dominant ideas of bucolic ruralism as the bedrock of English values. However, this is not unique to the twenty-first century. In *Electric Eden*, Rob Young charts a lesser-trodden path through twentieth-century popular folk music highlighting an alternative and experimental English rurality subsumed within dominant mythologies of the idyllic countryside. As he describes, as the 1960s counterculture burnt out, acts including Vashti Bunyan, The Incredible String Band, Mark Bolan and Nick Drake searched for a more inclusive sense of English identity in their music, embracing the dark history of the countryside while folding psychedelics, acid folk and Eastern symbolism into their reimagining of rurality.[43] This music revelled in the rural roots of the magical, occult and pagan, themes often written out of the official histories of the countryside. For Young, this rural revisionism reached its apogee in the work of the 'modern antiquarian' Julian Cope, who fused paganism, Krautrock and hallucinatory psychedelia in a series of albums in the 1980s and 1990s that imagined him as a modern-day William Blake, surveying rural England from the top of Silbury Hill, invoking Odin while lamenting the death of pagan matriarchy.

Listening to Landscape: Hauntology and the Echoes of Albion takes up the story that Young leaves in 2010, considering how ideas of the English landscape have continued to be reworked in contemporary music, particularly noting the uptick in this work since the fateful Brexit vote in 2016. While classical, rock, country and folk music all remain important in the contested construction of English national identity, this book proposes it is the artists combining electronic instrumentation with elements of field recording, sampling and sound collage who responsible for much of this work, and are arguably producing the most incisive response to the question of Englishness through their engagement with landscape. These acts include Vic Mars, Craven Faults, Warrington-Runcorn New Town Development Plan, Megalithic Transport Network, Spaceship, Gilroy Mere, Bonfire Hill, Belbury Poly, Warmfield, Twelve Hour Foundation, The Heartwood Institute, Kemper Norton, epic45 and many, many others besides. These musicians span multiple genres – post-rock, dark ambient, modular electronic, folktronica, synthwave – but are united by an apparent preoccupation with summoning the essence of the post-pastoral English landscape, often working with related ideas of place identity, memory and loss.

Electronic music's 'landscape turn'

This book explores how ideas of England and Englishness are being reimagined through a contemporary 'soundtracking of the nation' emerging at the interstice of electronic music, field recording and site-specific composition. It's not a definitive overview of the musical acts inspired by, or working with, the English landscape in the twenty-first century. It doesn't seek to delimit any sort of emergent scene or define the boundaries of a putative electronic 'landscape' genre: the Select Discography at the end of the book is a sampler rather than an encyclopaedic listing. Notwithstanding this enforced selectivity, the book seeks to profile a range of contemporary musicians working today whose music seems to embody a *critical* disposition towards the English landscape by exploring the relation between its past and present. Here, the conceptual framework of *hauntology* is often crucial as a way of exploring the presence of absence. Rather than offering a mimetic interpretation of the landscape, the hauntology encourages a radically nostalgic engagement with place that shows how the landscapes of Englishness are layered with ghosts and memories: it is about reading the landscape as a palimpsest of ways of life that have been repressed, forgotten and obliterated, revealing the forces that created the duplicitous landscape of Englishness so beloved by the right-wing populists seeking to 'take back control' of their 'green and pleasant land'.

Contrary to the suggestion there is a singular English landscape whose characteristic physical and social forms can be enumerated, and celebrated, the acts considered in this book expose the mythical production of nationhood by moving beyond the idealization of England's rural idyll. While all find beauty, and inspiration, in the landscapes that surround them, they reject bucolic English rurality in favour of a more nuanced, and localized, engagement with the darker side of the landscape, revelling in the eerie and disquieting qualities of place. Yet their music also explores the weirdness of the English landscape too, exploring the rupturing presences that are profoundly 'out of place'.[44] Tellingly, these counter-topographies of Englishness often move beyond the rural downlands and villages of Southern England, with much of the music discussed in this book having been written, performed or produced in the suburbs of post-industrial Northern towns, in faded seaside towns or among the ruins of Brutalist housing estates. Many focus on what might be termed 'ordinary' landscapes: spaces which are far from iconic, such as the pylon-studded wastelands or edgelands that ring many towns and cities or the low-lying allotments and caravan parks that cling to the coast in areas experiencing erosion and sea level change. Few invoke pristine or wild nature: even when the musicians profiled concern themselves with areas with little human habitation, such as the moors and peaks of Northern England, they are often exhuming traces of forgotten human occupation. Ancient Neolithic trackways, standing stones and deserted medieval villages feature prominently, as do industrial ruins, abandoned quarries and nuclear bunkers – landscapes that buzz, hum and sing with history.

This book hence explores how these places are summoned in the work of select contemporary electronic musicians, examining the relationship between the 'sonic landscapes' they are producing and the contested nature of contemporary Englishness. It does this in the context of the wider twenty-first century interest in folklore and myth that has bought sometimes arcane interests in landscape into the mainstream. This has perhaps been noted more in relation to film, literature and TV than music, with the films of Mark Jenkin and Ben Wheatley, the novels of Benjamin Myers, Andrew Michael Hurley and Fiona Mozley, Mackenzie Crook's slow-burning *Detectorists,* Charlie Cooper's *Myth Country* and Sally Wainwright's Yorkshire-based dramas exemplary of this resurgent interest in the traces of the past.[45] None offers a conventional or idyllic view of the English landscape: instead, they dig beneath the surface of the rural idyll to expose hidden histories of exploitation, exclusion and expulsion. They contest univocal conceptions of Englishness, reclaiming the landscape in the name of the local, the progressive and the environmental, wrenching these away from the provenance of nativist nationalism.

Listening to Landscape explores how the English landscape is represented not in literature, film or TV, but in experimental and electronic music. In so doing, it reflects on the relations of Englishness, place and belonging in the post-Brexit era, noting hauntological electronic music's ability to evoke the atmosphere and texture of landscape in ways that disturb existing narratives of place. Throughout, debates on soundscapes and the acoustic ecology of place are noted, especially as these relate to traditions of field recording, but the emphasis is more on the way that artists layer electronic sounds, instrumentation and samples to connote the present and past of the landscape. Exploring this sonic splicing of time and space, the book argues that music can be a particularly powerful medium for revealing obscure and forgotten histories of inhabitation, with the landscape-inspired artists on a diverse range of labels, including Wayside & Woodland, Ghost Box Records, Woodford Halse, Castles in Space, Modern Aviation, Plenty Wenlock, Analog Trash and Clay Pipe Music, to mention but a few, sharing an apparent preoccupation with disappearing, forgotten or threatened English landscapes. Often, the resulting music is deeply 'uneasy listening', emphasizing the weird and eerie, the uncanny and the repressed.[46]

The following pages explore the cultural significance of this music. Noting it characteristically works with a metaphysics of absence and presence that makes English landscapes both familiar and strange, and the next chapter details how this music is influenced by the critical theory of hauntology, to the extent some have even talked of the emergence of 'hauntronica' as a specific genre. This chapter considers the music that has been cited as exemplary of this genre and describes how it constructs an auditory *spectro-geography* that folds different times and places together to pose questions about landscape. It shows that hauntology's interest in ruination, obsolescence and horror overlaps with critical academic perspectives on landscape, where it has long been argued that landscape is duplicitous, obscuring the power relations that produce it by hiding histories of exploitation.[47] But while many academics study the representation of landscape at a particular historical juncture, hauntology involves a temporal enfolding that searches for anticipation of the future in the past and, simultaneously, the absence of these futures in the now. It exhumes the ghosts whose memories shape the present at the same time it ponders the weird presences in the landscape which, by all accounts, should no longer exist.[48]

This book therefore begins by discussing those acts – most notably those on the Ghost Box Records roster – who perhaps best exemplify this hauntological disposition. For many commentators, these bands – Belbury Poly, The Focus Group and The Advisory Circle in particular – are defined by their particular and sometimes *peculiar* attitude to the landscape. Using analog technologies, samples and knowing references to the TV soundtracks, Radiophonic experimentation and public information films of the 1970s, these acts construct a *heteroglossia* of musical reference points from both the recent and distant past, producing music that melds psychedelia, electronic and folk influences in its summoning of place and time. Despite many of these musical influences being familiar to listeners, especially those from the 'haunted generation' that grew up in the 1970s, the resulting music is far from comforting.

Having established the parameters of hauntology as a *critical* creative practice, the book moves across different English cities and regions – from the Lancashire New Towns to the Suffolk coast, from Cornish stone circles to the fens of Cambridgeshire, from Salisbury Plain to the peaks of the Pennines – showing that in every corner of the country there are musicians working with landscape and hauntology in inventive and original ways. In each case, the book considers how these engagements contribute to our understanding of England and Englishness in the post-Brexit era. Along the way, it explores how contemporary musicians are engaging with the utopian uncanny, are finding nature in the urban margins, are perambulating the gothic coast, walking weirdly into the countryside, and digging down deep into England's past. As we will see, they are finding echoes of Albion everywhere.

Further Reading

Ashbridge, Chloe (2023). *Rewriting the North: Contemporary British fiction and the cultural politics of devolution*. Taylor & Francis.

Featherstone, Simon (2009). *Englishness: Twentieth-century popular culture and the forming of English identity*. Edinburgh University Press.

Kenny, Michael (2024). *Fractured Union: Politics, sovereignty and the fight to save the UK*. Oxford University Press.

Masud, Noreen (2023). *A Flat Place: Moving through empty landscapes, naming complex trauma*. Melville House.

Matless, David (2023). *About England*. Reaktion.

Niven, Alex (2019). *New Model Island: How to build a radical culture beyond the idea of England*. Watkins.

Niven, Alex (2023). *The North Will Rise Again: In search of the future in Northern Heartlands*. Bloomsbury.

Young, Rob (2011). *Electric Eden: Unearthing Britain's visionary music*. Macmillan.

Ware, Vron (2022). *Return of a Native: Learning from the land*. Watkins.

2

The Past is Happening Now

Avebury, Wiltshire, has held fascination for generations of antiquarians, historians and archaeologists. A village situated within the largest Neolithic henge in the world, it's known through a bewildering range of place myths and superstitions, including associations with satanism that alarmed the church in the Middle Ages. As such, it's perhaps unsurprising its stones have long attracted artists, truth-seekers and savants. Paul Nash, an artist drawn to landscapes that spoke to his preoccupation with mortality, visited them in 1933, finding them 'wonderful and disquieting'; his contemporary John Piper rendered Avebury in watercolour, gouache and collage, with fragments of crumpled paper capturing the textures and patterns of the stones in an expressive, energetic composition entitled 'Archaeological Wiltshire' (1936). In the 1970s, Derek Jarman produced one of his first films there: *A Journey to Avebury* (1971) was an impressionistic home movie hinting at the psychic 'pull' of the stones, suggesting their indecipherable nature has invited acts of pilgrimage and veneration down the ages.[1] Ben Edge's (2024) more recent documentary film *Avebury* likewise devotes attention to the pagan communities who live and practice within the Neolithic henge.

Avebury also figures widely in fiction, an enigmatic setting lending verisimilitude to horror and fantasy alike. Renamed Milbury, it became the setting of Trevor Burnham and Trevor Ray's *Children of the Stones,* released to coincide with the HTV series of the same name in 1977. The complex, time-shifting plot suggested the Stones were acting as conduits to the negative power of the universe, forcing the village's inhabitants to endlessly replay the past as they become ensnared in multiple time-loops. Describing four time-loops, *Children of the Stones* depicted Milbury as endlessly playing out the events of the first loop in which a druidic priest initially harnessed the power of the stones. Combining sci-fi and supernatural themes, *Children of the Stones* was notoriously among the scariest TV programmes ever produced for young viewers, with Sidney Sager's memorable soundtrack combining eerie choral chanting, ethereal polyphonic harmonies and rumbling drum rolls to create a pronounced sense of foreboding.

In 2009, Belbury Poly released *From an Ancient Star* on Ghost Box Records. The front cover is of the stones bathed in a crepuscular, purple light: megaliths caught in the light of an intense star, or perhaps that of a visiting spaceship (Figure 2.1). The rear cover continues the theme with mysterious lights above a stone circle, while the CD booklet appears to mimic a 1970s *Fortean Times*-style magazine of the unexplained, promising stories of alien astronauts and portals to other dimensions. One advertised article on the relations of mathematics and folklore is credited to 'Aubrey Hutton', a combination of the names of British archaeologist Aubrey Burl and historian Ronald Hutton, both regarded authorities on Avebury's prehistory. *From an Ancient Star* hence makes connections between the astronomical and the ancient, tapping into the science fiction narratives of *Children of the Stones*, as well as the madcap ideas of Eric von Däniken, briefly popular in the 1970s, particularly the theory that ancient

Figure 2.1. *From an Ancient Star.* Courtesy: Julian House.

civilizations were shaped by extraterrestrial technology. Yet alongside this, the CD's *faux* contents page promotes a feature on model villages of the type made by railway enthusiasts, perhaps hinting at the idea that we all live in some sort of fake reality, manipulated by forces beyond our comprehension.

UFO sightings around ancient megaliths? Portals to other dimensions? Miniature worlds? The music certainly reflects this weirdness. Some tracks sound like the theme to 1970s science programmes, others offer synth-propelled, throbbing disco in the style of Giorgio Moroder. Opening track 'The Hidden Door' has a queasy melody that leads us out into 'the unexplored blackness of outer space', while 'arpeggio synthesizers suggest a phalanx of stars streaming past on either side'.[2] 'A Year and a Day' starts with gentle, folk-like guitar before quizzical synth notes drift in, a discombobulated woman's voice heard singing indistinctly in the background. 'The All at Once Club' sounds like The Beatles' 'Get Back' played on the cheesiest fairground organ, vocoder effects adding a further layer of 1970s verisimilitude; 'Time Scale' samples Elizabethan poet Samuel Daniel's poem 'Delia' (*Time, cruel time /come and subdue that brow*) over sparse percussive instrumentation suggesting the ticking of a clock. 'Widdershins' builds a simple medieval-sounding recorder melody into prog-era cacophony, while 'A Big Day Out' is a dub excursion.

Jim Jupp, the co-founder of Ghost Box Records who records as Belbury Poly, suggests this musical collage emerged from a realization that an early version of the track 'From an Ancient Star' felt like 'the soundtrack to a mysterious world-type documentary programme of the 1970s' as well as 'cosmic disco from the same period'.[3] He recalls that the idea of humanity being dwarfed by cosmic powers was drawn from the work of H. P. Lovecraft, the record reflecting 'a tradition of British science fiction where you've got on the one hand a very traditional background, with very ancient things, but you've got this weird, cosmic stuff happening'.[4] The album unquestionably captures this sense of temporal dissonance, moving between the future and the past, sometimes in the same track: one features a 1960s sample of a folk song layered beneath space-age synth. As Jupp explained in an interview with Mark Fisher:

> One of the themes that crept into the record, and that has obsessed me personally for a long time, is the idea of eternalism and non-existence of time. It's the notion that everything that has happened and will happen and all parallel world outcomes are superimposed in one block time . . . we've always imagined that the Ghost Box world is a kind of an 'all at once' place where all of the popular

> culture from 1958 to 1978 is somehow happening all at the same time. I think this is one of the reasons why each release we put out looks like it comes from a particular moment in this period and yet can reference much earlier and later events at the same time.[5]

Here, the look and feel of the release, the borrowing of disco motifs, the influence of the *Children of the Stones* soundtrack, and a fixation with cosmic conspiracy theories all locate *From an Ancient Star* in the late 1970s. But this is a twenty-first-century facsimile of 1970s music, one that summons the *atmosphere* of the times by emphasizing and exaggerating some of the era's fixations with temporality, the cosmic and the unexplained. As Alexis Petridis noted in his review in *The Guardian*, even when stripped of its carefully designed packaging, *From an Ancient Star* still succeeds in creating 'a strange, spellbinding imaginary world'.[6] This is the weird world of Ghost Box Records, the record label that has become synonymous with hauntology and 'things half-remembered from a fictional past or a parallel world'.[7]

Defining the hauntological

Hauntology has been something of a twenty-first-century preoccupation in critical cultural theory, inspiring literature, art, photography and film as well as electronic and experimental music. Though hauntology is a contested, and sometimes even reviled, term, it generally describes how the past circulates in the present as a spectral and uncanny presence that needs to be exorcized to make sense of the now.[8] The term has its origins in Jacques Derrida's (1992) lectures on the 'Spectres of Marx', a riposte to those who claimed the fall of the Soviet Bloc in the late 1980s heralded the 'end of history'. Contra those on the right arguing that the collapse of communism marked the final triumph of neoliberalism, successfully wedding capitalist systems of production and consumption to ideas of liberal democracy, Derrida pointed out that never in human history had so many been subjugated, starved or exterminated. Picking over the bones of Marxist theory, Derrida argued for a radical, new 'internationalism' to support those alienated by this new, rapacious capitalism.

While the term hauntology only occurs fleetingly in Derrida's post-mortem of Marx's canon, it has become a key analytic in contemporary cultural theory, with multiple commentators taking up Derrida's challenge of 'letting ghosts speak'. Geographers, historians and anthropologists, among others, have subsequently addressed questions of trauma and haunting, often exploring the wounds of colonial capitalism, war and suffering, but other times exploring the more mundane ways the past is banished from the present.

Hauntological work on place has, for example, latterly explored the significance of empty shops on British high streets,[9] the loss of community in the face of gentrification[10] and the voids where industry used to be.[11] In each case, hauntological inquiry raises questions about what has been lost and how such disappearance limits our understanding of what might have been (working from the definition of hauntology as an 'ontology of absence').[12]

Despite this inter-disciplinary diffusion, hauntology is a term now indelibly connected with the work and legacy of Mark Fisher. A philosophy lecturer who attracted considerable attention through his k-punk blog, established in 2003, Fisher was adept at showing how French post-structuralism and critical theory could be applied to film, art, literature and music, as well as to the landscape itself. Tellingly, some of the last entries on the k-punk blog were about his own engagement with the landscape of Suffolk, the county he came to call home. Made with sound artist Justin Barton, *On Vanishing Land* (2013) was a hallucinatory audio essay of a walk along the Suffolk coastline, infused with the melancholia that W. G. Sebald also experienced on that very same coast, as documented his 1995 book *The Rings of Saturn*. For Fisher, Suffolk's near-deserted heaths and beaches were inherently hauntological, replete with sinister and forgotten presences that could be divined only through an occult cartography.

On Vanishing Land ends at the eerie serenity of Sutton Hoo's Anglo-Saxon cemeteries but starts near Fisher's home at Felixstowe, close to the massive container port he described as a 'nerve ganglion of capitalism'. Speculating the ships on the horizon were bringing in medical and pharmaceutical supplies, he ruminated on the seeming 'pandemic' of mental health issues in society. This interest in mental health was among Fisher's main preoccupations, one of his central ideas being that the pharmaceutical industry could not offer the solution to depression and anxiety. Instead, he argued that alienation was inherent to neoliberal society, the result of economic and political processes pushing people into precarious and unrewarding jobs. For Fisher, depression was the 'shadow side' of the entrepreneurial culture that triumphed under Thatcher, who promoted the neoliberal doctrine that everyone could make it in society if they tried.[13] Suggesting there was nothing inevitable about neoliberal capitalism's colonization of everyday life, Fisher was distraught at the seeming absence of any feasible alternative to the capitalist order.

Fisher provided a unique reading of the 'crisis of the Left' in the twenty-first century, drawing on Benjamin, Freud and Derrida to grapple with the seemingly pathological impasses of retrogressive postmodern culture. His 2009 book *Capitalist Realism: Is There No Alternative?* likewise borrowed liberally from Fredric Jameson and Slavoj Žižek to explore why it was seemingly easier to imagine the end of the world than the end of capitalism. In this book, he highlighted the pervasive sense of exhaustion on the left and bemoaned the impossibility of articulating a critique of capitalism that was not immediately dismissed, recuperated or commodified. Fisher frequently illustrated this argument with reference to popular music: as he described it, while the music of the 1960s and 1970s was often revolutionary, in the twenty-first century, pop music sounded tired, recycled and in the grip of styles established long ago. So, despite the crises of global capitalism that deepened exploitation and precarity, Fisher suggested popular culture seemed unable to express any sort of alternative to the advanced capitalist notion of competition as virtuous. But exploring *unpopular* culture, Fisher found some music that expressed the melancholia of neoliberal Modernism almost perfectly, finding hauntology in the music of Burial, The Caretaker, Philip Jeck and Mordant Music, among others. While these acts were not connected through any semblance of a scene, Fisher suggested they exhibited a particular sensibility – 'an existential orientation' – towards the past, their music somehow out of synch with the present. In his essay 'The Slow Cancellation of the Future', Fisher also explored how their use of analogue technologies – old synths, cassettes, vinyl, videotapes – communicated this sense of overwhelming melancholy.[14] As he described, hearing the deterioration of material artefacts – the hiss of a tape cassette, for instance – returned the listener to the era of popular and progressive Modernism this technology was associated with: in England this was the post-war era of Penguin paperbacks, brutalist architecture and the BBC Radiophonic Workshop. For Fisher, hauntological music asked us to consider why the conditions that allowed these radical cultural forms to emerge in the 1960s and 1970s no longer exist. Fisher's thesis was that we might find an exit from the 'closed horizons' of capitalist realism by returning to the 'lost futures' of the past, particularly those of the post-war years when the welfare state was founded and social democratization set in motion.

Addressing the ghosts of our lives, hauntology can be seen as a form of radical nostalgia. It's a critical theory that has also become an aesthetic methodology in visual art, literature and music. Yet the latter has embraced hauntology's fixation with exhuming the futures of the past to the extent that its key sonic gestures – the use of obsolete musical technologies, temporal disjuncture and retrofuturism – have sometimes become cultural cliché. The music of The Caretaker, for example – the alter ego of musician Leyland Kirby – is often cited as the quintessence of hauntology. Originally inspired by the haunted ballroom scene in *The Shining*, Kirby's Caretaker project samples old, crackly shellac 78s, smothers them in reverb and oftentimes lets them dissolve in a bewildering audio fog. The Caretaker's 2006 album *Theoretically Pure Anterograde Amnesia* tellingly referred to a medical condition whereby an individual is unable to make new memories. As Mark Fisher noted in his sleeve notes to this album, this encapsulated the postmodern condition whereby it appeared increasingly impossible to envision the future, with society locked into 'endless rewind'. For Fisher, The Caretaker's work offered a critical diagnosis of society's fixation with nostalgia, suggesting this was endemic to capitalist realism. Foregrounding questions of temporality, The Caretaker showed time itself was 'out of joint', Fisher describing *Theoretically Pure Anterograde Amnesia* as a depiction of the 'wounds of time'.[15]

But when Kirby pushed The Caretaker project further, into an exploration of dementia across the six album *Everywhere at the End of Time,* it became an unlikely musical meme. Each LP was more washed out and confused than the last, suggesting the different stages of a descent into the later stages of Alzheimer's: listening to all six LPs in one sitting became a challenge taken up by teenagers who proclaimed it 'the scariest music ever'. Formulaic copies of the albums followed as fans used online software to knit together slowed-down samples of big band tunes to provide their own takes on the sounds of dementia and loneliness. At the same time, the Caretaker's music became the soundtrack to Tik Tok videos and subreddit gifs depicting AI-generated 'backrooms' – liminal spaces including empty malls, deserted children's parties and yellowing hotel rooms – designed to provoke a sense of anxiety. So while The Caretaker's experimental commentary on memory and loss was praised by dementia charities, the decomposing melodies that characterized his work were easily co-opted, becoming monetized online content. What Fisher had described as a music that resolutely refused to 'give up on the future' was transformed into yet another commodified symbol of generational angst.

Presenting . . . Ghost Box Records

The Caretaker project finished in 2019, with Kirby's final releases under that identity dedicated to Fisher, who committed suicide in 2017, aged forty-eight. However, the record label Ghost Box Records, described by Fisher as providing 'uneasy listening to the letter', continues to operate, many of its acts having been central in defining what has sometimes been termed 'hauntronica' – a woozy electronic invocation of lost futures that has become perhaps the most obvious musical expression of hauntology. Launched in 2004, Ghost Box's earliest releases were diverse, spanning the sinister Lovecraftian weirdness of Eric Zann (actually Ghost Box co-founder Jim Jupp) and Mount Vernon Arts Lab (Drew Mulholland) through to the quirky sample-based playfulness of The Advisory Circle (Cate Brooks) and The Focus Group (Julian House). Yet despite this variety, Fisher seized upon the label's musical and aesthetic identity as the quintessence of hauntology, not least its citation of hazily recalled 1970s children's TV programmes, public information films and late-night Open University broadcasts:

> The affect produced by Ghost Box's releases (sound *and* images, the latter absolutely integral) are the direct inverse of irritating postmodern citation-blitz. The mark of the postmodern is the extirpation of the uncanny, the replacing of the unheimlich tingle of unknowingness with a cocksure knowingness and hyper-awareness. Ghost Box, by contrast, is a conspiracy of the half-forgotten, the poorly remembered and the confabulated . . . Ghost Box releases conjure a sense of *artificial* déjà vu, where you are duped into thinking that what you are hearing has its origin somewhere in the late 60s or early 70s: not false, but simulated, memory. The spectres in Ghost Box's hauntology are the lost contexts which, we imagine, must have prompted the sounds we are hearing: forgotten programmes, un-commissioned series, pilots that were never followed-up.[16]

As Fisher noted, Ghost Box releases present as if they are the soundtrack to a programme recently recovered from the archive, or a compilation of clips from long-lost films. A particular preoccupation is the adoption of the aural and visual tropes of the public information leaflets and films of the 1960s and 1970s. The Focus Group, for example, incorporate children's exercise records, library sound effects and public information films in their aural recollection of ways the post-war state attempted to produce a heathy, happy citizenry. On *Hey Let Loose Your Love*, 2005, and *We Are All Pan's People*, 2007, longer tracks are sandwiched between shorter 'logotones' or idents inspired by TV continuity announcements, as if one is flicking channels between school's television, daytime news coverage and an instalment of *Top of The Pops* in the prog rock era.

But there are also hints on these records that young people were exposed to more harrowing imagery, including campaigns warning of the threat of rabies, the dangers of climbing into electrical substations, the risks of playing near railway lines. For example, in government missives of the time, little distinction was made between the ambient threat of nuclear war and the banal terrors to be found on a farmyard in the English countryside, replete with murderous slurry pits and deadly farm machinery (as portrayed in the traumatic 1977 Central Office of Information film *Apaches*). The Advisory Circle sampled these coldly authoritarian public information films on *Mind How You Go* (2005), drawing out their disturbing undercurrents while noting their stylistic and aesthetic continuity with children's television of the period, suggesting even teatime viewing was infused with deadly foreboding. Cate Brooks (of The Advisory Circle) recalls their impact on her:

> I grew up amongst television programmes such as *Threads* and documentaries about nuclear war, or at least the threat thereof. Another very chilling extension to this was finding 'Protect and Survive' leaflets in public libraries, amongst all the other Central Office of Information [*leaflets*] for things like Child Benefit and Rent Allowance. Finding and reading those leaflets in such suitably 'cold', subdued, and institutional surroundings was a particularly formative touchstone for The Advisory Circle.[17]

Hence, while much of the music on Ghost Box Records seems naggingly familiar to those who grew up in the 1970s, it's not easy listening, frequently alluding to the dark sides of Modernity that irradiated everyday life, not least the existential threat which the haunted generation lived with: mutually assured nuclear destruction.

While Ghost Box Records has sometimes been labelled a whimsical exercise in *faux* nostalgia, for Fisher it represented a critical attempt to question the lost futures of the past, misremembered and otherwise. The fact much of Ghost Box's output summoned the post-war years of popular Modernism, when the future seemed up for grabs, seemed to pose important questions about what the future could – and should – have been. As Fisher notes, the label's strong visual aesthetic was central to this: the album covers, designed by Julian House (who founded Ghost Box with Jim Jupp), evoked a post-war public sphere where school textbooks and public service manuals had the same aesthetic status as mass-produced Penguin book covers or TV idents. With their clean lines and distinctive logo, Ghost Box's initial releases possessed a distinctly modern identity influenced by British Rail Corporate Identity manuals of the 1960s, library music album covers and, most obviously, Germano Facetti's designs for the mass-produced Pelican paperbacks that provided 'intelligent books' at affordable prices (Figure 2.2).

For Mark Fisher, the 'Britishness' of these varied reference points was significant, indicative of twenty-first-century yearnings for the public service broadcasting tradition of the post-war years when popular culture was both challenging and experimental.[18] This witnessed Ghost Box acts citing, and sampling, the soundtracks of TV series that inserted the *avant garde* of electronic music into the mainstream in the 1960s and 1970s. A touchstone was the work of the Radiophonic Workshop, the BBC sound effects unit that pioneered new methods of tape manipulation and electronic oscillation to add music and sound effects to schools' programmes, regional news broadcasts and radio dramas from the late 1950s onwards. Disbanded in the 1990s, the workshop is perhaps best remembered for providing incidental music and effects for popular science fiction series, starting with Nigel Kneale's *Quatermass and the Pit* (1959), reputedly the first television production to incorporate electronic music. Drew Mulholland's 2007 release on Ghost Box Records (as Mount Vernon Arts Lab), *The Séance at Hobs Lane*, directly cites *Quatermass and the Pit*, Hobs Lane being the site of the numerous apparitions that the eponymous Professor Quatermass ultimately deduces are alien-induced psychic projections, poltergeists that are traces of a shared racial memory.

Film scholar Adam Scovell argues that *Quatermass and the Pit* gives Mark Fisher's hauntology a 'dramatic corporality' by suggesting the repressed remains active even when it appears dead and buried.[19] But in contrast to much of the folk horror that Scovell discusses, Kneale's work focused on the psychic rather than physical manifestation of the past and eschewed body horror in favour of something far more unsettling (in this case, the idea that the human race is evolved from the alien). Fisher himself argued that Kneale's work questions the boundaries between the scientific and the supernatural, leading us to the ultimate realization that the 'world in which we live is profoundly more alien and strange than we had previously imagined'.[20] For Fisher, the idea that cultural memories can be understood *archaeologically*, as relics of previous traumatic events, tied hauntology into the notion that human societies are destined to senselessly repeat

Figure 2.2 Ghost Box Records Night flyer. Courtesy of Julian House.

past events (hence his insistence that we are enmeshed in mythic structures that pre-date us and which we cannot live outside). Fisher insisted time itself is far from smooth, being 'broken' into traumatic moments, with the Quatermass series exploring the way that the deep past can still inform the present.

The work of Nigel Kneale also looms large over Eric Zann's 2005 *Ourobordindra* release on Ghost Box, inspired by Kneale's 1972 *The Stone Tape* TV play, first broadcast at Christmas on BBC2.[21] This concerns a group of scientists setting up an experimental sound facility in a Victorian mansion only to discover that the very stones of the building have acted as a recording medium going back centuries. The lead scientist Peter Brock's words are quoted on the back of the release: 'Don't you get it yet? It must work like a recording. Fixed in the floor and the walls, right in the substance of them. A trace of what happened in there. And we pick it up. We act as detectors – decoders – amplifiers'. Zann's album is an anxiety-inducing ambient drone overlaid with indistinct, overlapping voices, twanging sitars and oscillating sound effects. Like the TV series, it's also punctuated by moments of real dread. The final track, 'The Human Chord', is a taut, howling swirl of descending chords and echoing human voices, mirroring the conclusion of the play where Jill Greely (played by Jane Asher) falls back in time, and to her death. The album ends in a noisy cacophony before a final faint 'amen' is heard, and we are left with the sound of carrion crows, ancient harbingers of death.

Fisher highlighted *The Stone Tape* as an exemplary exploration of the 'terrible compulsion to repeat', noting the impossibility of creating societies that can make sense of their existence without any reference to the wounds of the past.[22] In this sense, Kneale's sometimes-disturbing exploration of residual haunting clearly resonates with many of the conceits of hauntology. Perhaps unsurprisingly, *The Stone Tape* has also been revisited by others outside the Ghost Box roster, with a 2015 adaptation for radio by Peter Strickland and a 2023 tribute on Hidden Britain tapes featuring Drew Mulholland alongside The Hardy Tree, E. L. Heath, The Heartwood Institute and other notable hauntronic acts. However, the original soundtrack, by Desmond Briscoe and Glynis Jones of the Radiophonic Workshop, was not issued until 2019, released on bilious green, fluorescent vinyl to mirror the luminous display of the scientists' sonic oscilloscope in the play. Listening to this soundtrack, the almost perfect synergy between Kneale's time-shifting parables of the uncanny and the other-worldly sounds of the Radiophonic Workshop becomes clear: Briscoe and Jones' sparse soundtrack is every bit as disconcerting as Zann's twenty-first century reimagining, all babbling electronic pulses, slashing electronic effects and treated voices. Its main motif is a descending melody played on a squelching, atonal keyboard that creates an eerie, unsettled atmosphere emphasizing the strange goings-on in the research building.[23]

Yet for many from the 'haunted generation' that grew up in the 1970s and 1980s, the Radiophonic Workshop remain most associated with *Doctor Who*. Delia Derbyshire's futuristic theme, created from tape manipulation of plucked piano strings and clanging metal, heralded the arrival of the series in 1963, with the Radiophonic Workshop stepping up to provide much of the incidental music from the 1970s onwards: Malcolm Clarke's soundtrack to *The Sea Devils* (1972) was particularly notable, an obtrusive electronic babble composed on an EMS Synthi 100 (Delaware) synthesizer, frantic stabs of electronic noise quite unlike anything heard on television before or since. This was in a period when *Doctor Who* was earth-bound, Jon Pertwee playing a Quatermass-type authoritarian figure grappling with mad scientists and alien species within the everyday infrastructures of power stations, coal mines and chemical works. Yet the writers of *Doctor Who* were, like Kneale, often concerned with the reverberations of the past: in *The Sea Devils* the alien species threatening the South coast of England were actually a primordial aquatic species whose name was retained as an ancient race memory, while in *Inferno* (1970) experimental drilling releases a liquid that transforms those who come into contact with it into primeval *homo sapiens* poorly acclimatized to the earth's temperature in the present. *The Daemons* (1971) memorably dabbled with the tropes of folk horror, with the unsealing of a burial mound allowing the lone survivor of an alien race to resurface, a psychological monstrosity that threatens local villagers. Other Doctor Who episodes, including *The Mind of Evil* (1971), tapped into the nuclear anxiety of the 1970s, suggesting modern science often walked the fine line between progress and destruction, a theme explored in the *Claws of Axos* (1971) when alien Axons threaten to turn Dungeness power station on the coast of Kent into a gigantic nuclear bomb.

Doctor Who's concern with time-travel, foolhardy scientific experimentation and the collision of rationality with superstition resonates with many of the fixations of hauntology, underlining why the Radiophonic Workshop's music has proved quite so influential for those on Ghost Box. Belbury Poly's (2020) *The Gone Away*, for example, takes its cue from Paddy Kingsland's incidental music from 1980s *Doctor Who*, all flute melodies and quizzical synthesizer refrains, while The Advisory Circle's (2021) *Full Circle* even takes its name from a *Doctor Who* episode exploring evolution and regression, seemingly using this as an analogy for the recuperation of late twentieth century culture, art and architecture. While other lysergic television that aired in the 1970s – *The Owl Service* (1969), *Tomorrow People* (1973) and *Sapphire and Steel* (1979–82) – clearly influenced Ghost Box acts, *Doctor Who* is the most direct progenitor of its hauntronica. However, the influence of the Radiophonic Workshop which irradiates Ghost Box releases in the twenty-first century can also be found in the work of a handful of acts that emerged the early 1990s, a full decade before the term hauntology had been used to delineate a particular musical style or genre. Here, it was Broadcast, alongside other Birmingham-based electronic acts including Plone, Pram and l'Augmentation, who pioneered a retro-futurist sound directly influenced by the Radiophonic Workshop, particularly the 1960s *music concrète* and tape manipulation of Delia Derbyshire, John Baker and Daphne Oram.

Once upon a time in the West Midlands

Birmingham in the 1990s was the second biggest city in the UK but lagged badly behind many other cities undergoing a cultural revival underpinned by popular music and the creative industries. Manchester had spawned Madchester, The Stone Roses and a burgeoning club scene centred on The Haçienda; Sheffield was making plans for a National Centre for Popular Music as Pulp scaled the charts; Glasgow had been European City of Culture in 1990; Bristol had trip hop. Even Liverpool had begun to reverse a decade of political turmoil and population loss, with the redevelopment of Albert Dock, the arrival of Richard and Judy, The Farm and The La's. Birmingham, on the other hand, punched well below its weight, the only real shining light being Simon Rattle's Symphony Orchestra.

A few decades earlier it had been very different. Birmingham had long been an industrious place, the 'city of a thousand trades', flourishing in the 1950s and 1960s thanks to the rapid growth of the automobile industry. This prosperity was mirrored in the dramatic remodelling of the city on transatlantic lines, the city transformed into a futuristic urban ensemble of soaring flyovers, high-rise towers and pedestrianized precincts. The Bull Ring, the first indoor shopping centre in the UK, opened in 1964, and Herbert Manzoni's ambitious ring road was completed in 1971. Spaghetti Junction, opened in 1972, positioned Birmingham at the heart of the national motorway network, quickly becoming symbolic of a city whose fortunes were umbilically tied to that of the British motor industry. Birmingham lived up to its civic motto 'Forward': it was bold, innovative and distinctly Modern.

Yet by the late 1980s, the city had experienced years of disinvestment, the collapse of the British motor industry, high unemployment and social malaise. The city's Modern architecture became widely mocked, the future King of England describing the city in 1989 as one built for motorcars rather than people, a 'vast concrete maze'. John Madin's brutalist central library, an elegant inverted-1970s ziggurat, was singled out for opprobrium as a place for 'burning books rather than reading them'.[24] Responding, the increasingly entrepreneurial City Council pumped public money into speculative flagship schemes, overwriting the city's Modern iconography with a placeless postmodernism intended to make it more attractive to inward investment.[25] While many other British cities incubated their cultural industries, Birmingham poured millions into an International Convention Centre in the hope that foreign business tourists would flock to the city's first and only five-star hotel. The failed 1992 Olympic bid and the short-lived Birmingham Super Prix suggested otherwise.

But at precisely the same moment the city's 1960s architecture disappeared beneath a new postmodern blandscape, a new, young, generation of Birmingham musicians were rediscovering the counterculture of the 1960s. They scoured the Rag Market in the shadow of the Bull Ring searching for vintage fashion and flipped through the racks at Reddington's Rare Records hunting for obscure discs to spin at local club nights. Some decided imitation was the best way forward: Ocean Colour Scene became phenomenally successful by mimicking the proto-Britpop of The Small Faces and the 1960s Mod bands, while The Sea Urchins fused garage rock and indie jangle, interspersing original material with covers of songs by 1960s acts like The Eyes and The Creation. Broadcast, on the other hand, formed when James Cargill and Trish Keenan bonded at Mark MacDonald's 'Sensateria' evening over their love for short-lived 1960s experimental rock band, The United States of America, whose work combined primitive synthesizers, sound collage and hard rock to beguiling effect. Lead singer of Broadcast, Trish Keenan, later recalled the appeal of this psychedelia:

> I'm not interested in the bubble poster trip, 'remember Woodstock' idea of the sixties. What carries over for me is the idea of psychedelia as a door through to another way of thinking about sound and song. Not a world only reachable by hallucinogens but obtainable by questioning what we think is real and right, by challenging the conventions of form and temper. Bands like The United States of America, White Noise, A To Austr and The Mesmerizing Eye all use audio collage, clashes of sound that work more in the way the mind works, the way life works, extreme juxtapositions of memories and heavy traffic noise say, or reading emails and wasps coming through the wind.[26]

Booked for their first John Peel session in 1996, Broadcast quickly developed a reputation for combining these psychedelic influences with jazzy new wave cinema soundtracks from Eastern Europe and nods towards Joe Meek and John Barry's 1960s studio experimentation.

But, as Cargill noted, it was the tape experimentation and musique concrète of the Radiophonic Workshop that gave Broadcast their signature sound:

> The Radiophonic Workshop seems to exist in a place outside any obvious musical association. To think of it as kitsch, nostalgic or ironic is missing the point. There is no ironic value for me. It's sometimes quite disturbing, possibly because of the simplistic melodic content coupled with unrecognisable sound palettes, mostly it seems so reflective of that period late 60s/early 70s Britain. I actually think that it's quite a dark and disturbing era and sometimes the more out of context or more folky or childlike the composition, the more unsettled and out of place you feel.[27]

Cargill claimed the music of the Radiophonic Workshop transcended dark, heavy and masculine rock and lighter, marketed pop, creating a liminality which made the band simultaneously accessible and unsettling. For example, their first single, 'The Accidentals', was built around a looped jazz sample from the 1967 Joseph Losey film *Accident* about a male Oxbridge lecturer obsessed with one of his female students: the film's sinister themes are echoed in the song's lyrics about the boughs of a tree breaking. Trish's cold, disinterested vocals float over the mystery below, as if Nico had been born in west Birmingham rather than Köln. It was an astounding debut single, sounding both futuristic and retrogressive at the same time.

Like fellow retro-futurists Pram, who toyed with imagery from early cinema and surrealist art, and released an EP inspired by Alan Garner's *Owl Service*, Broadcast ultimately perfected a sound based on the familiar music of the past, theoretically accessible to mainstream audiences but always oddly disquieting, somehow out of time. This was reflected in their avant-pop image, a band stranded somewhere in the monochromatic 1960s: Trish wore white vintage, tunic dresses, James and the other members plain, buttoned-down shirts, a look somewhere between that of a lab technician and a preppie. On tour, they projected old industrial information films as they performed, offering their own brand of visual psychedelia (Figure 2.3). Record sleeves featured geometric forms, bold swathes of colour, Saul Bass typography, the band presenting an form of 1960s cool that had never really existed in the provincial Midlands. Their lyrics were often abstract, atmospheric and unsettling, suggesting a band self-consciously navigating the ruins of Modernity on its own terms.

Figure 2.3 Trish Keenan of Broadcast, 2010. Courtesy: Sam Ashlee.

Broadcast hence offered an exploratory prism through which the futures of the past could be reassessed, its failures evaluated. Here, architectural historian Owen Hatherley discerns an important connection between Birmingham's changing urban landscape and Broadcast's music:

> In Birmingham the groups I felt made most sense of the city were late nineties bands like Broadcast and Pram who are both from there and have this idea of Birmingham as a somewhat futurist but also sad and provincial place. There's a lovely song of Broadcast's called 'Michael A Grammar' about towers being knocked down in Chelmsley Wood [*outer Birmingham*]. It's a beautiful song that captures the air of Brum as being harsh and melancholic but also entirely ordinary, which I thought it did really well . . . Birmingham has always been . . . workaday, and I think that's reflected in the music.[28]

This sense of the city as a faded utopia, out of time with the late twentieth century and never quite reaching its potential, permeated Birmingham's retro-futurist scene. Broadcast were the acknowledged leaders of this short-lived scene, but several others contributed, as Hatherley notes. Pram, for instance, presented themselves as the ghosts of a fast-fading 'electrical-mechanical era' in an increasingly digital age (Figure 2.4). Their sometimes free-form post-rock combined unusual instruments like the Omnichord and theremin with trumpets, drums and toy keyboards: lyrically, they explored how exploration and travel defined Modernism, riffing on the memorialization of the space

Figure 2.4. Pram. Courtesy: Nick Sales.

race, polar expeditions and deep-sea exploration. Plone, meanwhile, tapped into the optimism of 1970s Schools programming, using analogue synths to create jaunty jingles and space-age pop that nonetheless rippled with unease (again, presaging Ghost Box Records, the label who ultimately put out their second, belated, LP *Puzzlewood*, in 2020). The short-lived l'Augmentation, on the other hand, seemed inspired by the work of Walter Benjamin, reimagining the streets of Birmingham as the Arcades of Paris, their solitary album *Pigalle* (1999) combining disconcerting fairground music and wistful jazz to produce a surreal soundtrack for perambulating the concrete underpasses of Birmingham. Through these arcane reference points, these bands offered something altogether more cerebral and interesting than the landfill Britpop that proliferated in the early 1990s, being infused with the melancholic existentialism Mark Fisher later claimed to be the essence of hauntology.

Yet Broadcast were the most celebrated among these acts, mainly because their music possessed an obvious commercial streak. Their second release – 'The Book Lovers' – was a minor pop masterpiece, combining '60s kitsch pop with edgy, atmospheric weirdness: it became a modest hit on the back of widespread radio play and the patronage of John Peel (who included them, alongside Plone, in his Meltdown festival in 1998). 'The Book Lovers' found its way onto the soundtrack of Mike Myers' (1997) *Austin Powers*, with the royalties helping Broadcast establish a recording facility in a unit in the old Bird's custard factory to the east of Birmingham city centre. It was here, in the white cube of a former industrial building, with echoing high ceilings, exposed beams and steel-framed Crittall windows, that the band recorded their much-delayed first album, *The Noise Made By People*. Released in 2000, it was feted in many quarters, including a gushing full-page review in *The Guardian*.

Successive albums attracted further praise, albeit over time Broadcast's sound became more minimal and monochrome, the pop edges replaced by a more abstract, dreamy surrealism. Tellingly, when Trish and James moved from Birmingham to Hungerford in 2009 – the scene of a notorious killing spree in 1987 by unemployed rural labourer Michael Ryan – their music shifted direction, displaying an increasing interest in folk horror, the arcane and the occult. They also began to actively collaborate with some of the artists on Ghost Box who had been inspired by them, making a 'Hammer horror dream collage' album with The Focus Group – *Broadcast and The Focus Group Investigate Witch Cults of the Radio Age* (2009) – that flipped incoherently between whimsical, folkish songs and snatches of phones ringing, music box chimes and fragments of earlier albums. Yet it was Broadcast's role in providing the score to Peter Strickland's (2013) *Berberian Sound Studio* that arguably cemented their status as the originators of a hauntronica whose influence can be heard now not only in experimental electronic music but across dream pop and chillwave through to the sophisticated psychedelic folk, kosmische and krautrock of acts including Death and Vanilla, Vanishing Twin and Jane Weaver. From modest beginnings in Birmingham, Broadcast somehow became one of the most important reference points in twenty-first-century alternative music. However, by the time *Berberian Sound Studio* appeared, Trish Keenan had died of complications connected to swine flu, aged just forty-two.

Haunted landscapes

An often-voiced concern about hauntology as a label is that it overburdens music with excess theoretical baggage, over-determining its meaning. There is perhaps some truth to this: attempts to summon the hauntological qualities of music can veer down obscure intellectual and cultural rabbit holes, imbuing songs with meanings never intended and influences that never were. Many acts described as hauntological disavow the label altogether. Irrespective, Mark Fisher's citation of Ghost Box Records as capturing the aesthetic and musical essence of hauntology, and, in turn, the label's obvious sonic debt to Broadcast's Radiophonica, provides a useful set of reference points for understanding the sonic parameters of hauntronica in the twenty-first century. As Fisher emphasized, hauntological music may sometimes sound 'ghostly', but its spectrality does not just refer to its evocation of spooky or weird atmospherics. Rather, it concerns the sonic articulation of a sense of loss and an acute awareness of the failure of contemporary culture to articulate any alternative to 'capitalist realism', especially through reference to the lost utopianism of post-war Modernism.

But what, precisely, does this have to do with Englishness and landscape? Geographer Callum Sutherland answers this question obliquely by suggesting that hauntology is about *displacement*: for example, the evisceration of place identity under globalization, the erasure of communities, the destabilization of Modern ideas of democracy and so on. He insists hauntology is of value as a critical lens on the past because it reconstructs the unexplored potentials of things that seem to have been exhausted or abandoned, stifled by the 'stale atmospheres of capitalist realism'.[29] Sutherland stresses the importance of what he terms hauntology's 'grotesque stratigraphy': an almost archaeological process of stripping back the layers beneath conventional or 'official' representations of landscape to explore the social, cultural and political detritus that lies beneath, fragments ignored by the heritage industry. In this way, hauntology disrupts capitalist hegemony, offering escape routes from contemporary capitalism.

Sutherland also stresses that hauntology encourages a consciousness of places and people that are physically distanced from us in both space and time, leading us to what might be surprising insights. Can we learn more about sustainable transport by looking at Neolithic trackways than smart motorways, for example? Should we revisit the often-derided hippy communes of the 1960s to appreciate how to have more environmentally friendly lives? Were forms of medieval customary law granting access to the commons preferable to contemporary forms of property rights? Sutherland's propositions suggest that hauntology offers a possible route to articulating postcapitalist desire in the contemporary moment, arguing that, for example, the countercultures of the 1960s and 1970s developed a green politics rooted in everyday sensibilities more suited to the climate emergency than any technical solution promoted by neoliberal capitalists in the present. In relation to landscape, this suggests the need for a hauntological reading that unearths the ways of life and 'structures of feeling' displaced in the name of capitalist progress. This is no straightforward landscape history: hauntology often demands consideration of that which appears tangential to the historical record, lacking any obvious connection to the present precisely because of its failure to embody capitalist desire. At times it veers on the surreal: like Dada, hauntology revels in the art of collage, adopting a non-linear perspective to show how the past persists in the present.

Here, the radical dimensions of hauntology's engagement with the landscape become clear: through tactics of dis/orientation, hauntology reveals the *psychic* scars left in the landscape when past forms of inhabitation were displaced in the name of capitalist progress. In this way, Oli Mould suggests it bears clear similarities to the urban psychogeography of the 1960s which used détournement and the *dérive* to see alternative spaces 'in-between, beyond and within the society of the spectacle'.[30] Yet what is perhaps different in hauntology is its desire not to escape capitalist space but capitalist *time*; it rediscovers disappeared Others who could not easily be calibrated with capitalist modernities, those deemed 'out of time'. Hauntology hence tangles the 'string of temporal linearity', folding different spaces and times together.[31] It shows a stone circle is just as haunted by the ghosts of a rural rave in the 1990s as by the echoes of ancient pagan rituals. Roman roads are juxtaposed with motorways, standing stones with power stations. An abandoned nuclear bunker and a deserted medieval village are granted the same ontological status. The ghosts of the premodern – witches, pagan worshippers, lepers, druids – manifest alongside the eccentrics, malcontents and travellers whose presence in today's landscape appears equally contested. Through a jumbled juxtaposition of old and new, hauntology denies the logics of accumulation through dispossession, the neoliberal formulation eviscerating the public realm and destroying the commons in the name of progress.[32]

Ghost Box Record's attempt to construct what Jamie Sexton terms a 'psychic heterotopia' encapsulates such landscape critique.[33] As Sexton notes, The Focus Group, The Advisory Circle and Belbury Poly collectively invoke an alternative British (or English) heritage incorporating spaces often condemned in the twenty-first century as 'failed' social experiments (e.g. post-war secondary schools, New Towns, municipal libraries). In turn, these are placed alongside 'valued' sites of English heritage, including relict forms from the premodern, such as megalithic stone circles, hill forts or chalk figures on hillsides. These are imaginatively located in the meta-fictional landscape of Belbury, a market town with a regular parish newsletter (which for a while was the de facto newsletter for the record label). According to Jim Jupp, the self-styled vicar of Belbury, 'It's a setting for a lot that happens in the Ghost Box world . . . It's not like it's *The Lord of the Rings*, and we're going to give you a map and a big book. But as we talk about it more, it makes more sense and . . . it's slowly developing into a fleshed-out world'.[34]

Despite its clearly fantastical elements, Belbury is very familiar to anyone who grew up in England in the post-war years:

> Belbury is at first an archetypal small English town, but what has become apparent is that there's no particular sense of it being anchored in a certain point in time. It's kind of an all-at-once place where it's simultaneously every year from 1960 to 1980 . . . In Belbury (like many real English small towns), this manifests itself in a jumble of architecture where mediaeval pubs nestle next to Brutalist concrete public libraries, or where a stone circle can be glimpsed from the roof of the multi-storey car park.[35]

The notion of Belbury as an unseemly collision of old and new is reproduced in the sleeve notes to Belbury Poly's *The Owl's Map* (2006) which talk of it being a market town on the border (presumably between England and Wales, where Jim Jupp and Julian House grew up). It's described as overlooked by the foreboding Iron Age ramparts of Belbury Hill, situated near a Neolithic Ring. Noting the town suffered in an opportunistic air raid in 1940, *The Owl's Map* sleeve also details its haunted manor house, the 'Polytechnic College', a Modernist public library and a 1960s estate pub, 'The Bury Bell'. Belbury is then a composite of different architectural styles and periods as found in many small English towns, albeit the mix of ancient and modern is disconcerting. The photos accompanying the text heighten this sense of unease, one being of a pair of hitchhikers caught blankly in the glare of car headlights, another the entrance hall of a stately home viewed via a mirror, a blurry figure at the top of the stairs suggesting ghostly apparitions.

This sense of things being a little awry in Belbury is developed elsewhere: for instance, *The Belbury Tales* (2012), the fourth album from Belbury Poly, includes a quote from Trish Keenan in the sleeve notes – 'It seems to me the past is happening now' – and is a loosely themed concept album which tells of a travelling salesman stopping off at *The Bury Bell* for a ploughman's lunch in the mid-1970s. Regaled by a local prog rock band, and under the influence of a spiked drink, he regresses into a befuddling world of medieval travellers (A Pilgrim's Path), Arthurian legend (Chapel Perilous) and devil-worshipping (Goat Foot) (Figure 2.5). As David Pattie notes, in *The Belbury Tales* the 'tropes and styles of seventies rural horror – the ominous chords, pounding rhythms, the grimly jaunty jigs and reels, the eerily melodic children's singing, the exaggeratedly rustic voices of the locals' combine to construct a 'seventies British folk–horror movie soundtrack'.[36] This expresses social conflict, 'articulating the pressures of multiple binaries: local and not-local, "them" and "us", Modernity and antiquity, progress and regression, urban and rural, nature and culture, civilization and wilderness, religion and enlightenment'.[37]

In this sense, the heady invocation of off-kilter rurality in *The Belbury Tales* emphasizes the uneasy ways in which post-war Modernity co-existed with surviving fragments of the 'dim and distant' past that were in fact not so far removed at all. This emphasizes that while we might be nostalgic for the imagined futures of the seemingly progressive 1960s and 1970s, the public cultures of this period were themselves haunted by the spectres of the past, ghosts kept alive via rituals that echo down the years: Morris dancing on village greens, partaking in a ploughman's lunch in a rural pub, folk music, the retelling of Arthurian legend and so on. Elodie Roy concludes that Ghost Box Records resists the idea of the past as a 'finished and frozen place', offering a subversive reading that shows it is always 'a changing country'.[38]

Partly because of the attention bestowed on them by Mark Fisher and partly because of their uniquely wonky curation of an alternative English heritage, Ghost Box Records have been repeatedly lauded by academics for their ability to 'melt past, present and future into an eerie hypnagogic coil'.[39] But as the record label moves beyond the twentieth anniversary of its first release, the meta-fiction of Belbury has been quietly pushed aside. Many of the latter releases on the

Figure 2.5. *The Belbury Tales*. Design: Julian House, Courtesy: Ghost Box.

label – the skewed acid rock of the Portuguese band Beautify Junkyards, for example, or the whimsical German synth pop of Sebastian Counts' ToiToiToi – have little to do with Belbury, or Englishness for that matter. But as Adam Scovell has noted, in the opening decades of the twenty-first century, there are 'a surprisingly large number of musicians churning out similarly haunting and disquieting sonic fictions that chime with the notion of an alternative Albion'.[40] It's these often unheralded musical acts that provide the focus of the following chapters, which move from Southern lowlands to Northern uplands, from rural to urban and from field to forest in the search for a music that captures the very heart and soul of English identity in the fractured post-Brexit era.

Further Reading

Christodoulou, Chris (2018). 'Haunted science: The BBC Radiophonic Workshop and the lost futures of hauntological music'. *Scene*, 6(2), 107–120.

Coverley, Merlin (2020). *Hauntology: Ghosts of futures past*. Oldcastle.

Donnelley, Kevin, & Molloghan, Aimee (eds) (2023). *Haunted Soundtracks: Audio-visual cultures of memory, landscape and sound*. Bloomsbury.

Fisher, Mark (2012). 'What is hauntology?' *Film Quarterly*, 66(1), 16–24.

Fisher, Mark (2014). *Ghosts of my Life*. Zero.

Fisher, Mark (2016). *The Weird and the Eerie*. Watkins Media.

Pattie, David (2016). 'Stone tapes: Ghost box, nostalgia and post war England'. In Whiteley, Sheila & Rambarran, Shara (eds), *The Oxford handbook of music and virtuality*, 392–408. Oxford University Press.

Reynolds, Simon (2011). *Retromania: Pop culture's addiction to its own past*. Macmillan.

Sexton, Jamie (2012). 'Weird Britain in exile: Ghost box, hauntology, and alternative heritage'. *Popular Music and Society*, 35(4), 561–584.

Sutherland, Callum (2023). 'Mark Fisher and reimagining postcapitalist geographies'. *Dialogues in Human Geography*, 13(1), 99–118.

3

Lost Landscapes of Modernity

Alongside Ghost Box Records, there is one other label often identified as central to the twenty-first century "landscape turn" in electronic and experimental music. Clay Pipe Music, curated and run by Frances Castle, has been responsible for over forty LPs since 2010, many displaying an acute interest in questions of Englishness, identity and place. Though less obviously preoccupied with folk horror or the occult than Ghost Box, many of these releases have alluded to the spectral and the absent presences in the landscape. Given this, it is perhaps unsurprising their roster has some overlap with that of Ghost Box, not least through the retro electronica of Cate Brooks, who records as The Advisory Circle for Ghost Box as well as under her own name for Clay Pipe. The label also put out the 2015 Welsh language collection *Friends & Enemies* by contemporary folk singer Sharron Kraus, who later collaborated with Belbury Poly on *Chanctonbury Rings*, a musical adaptation of a chapter of Justin P. Hopper's (2017) book *Old Weird Albion*.

But despite these overlaps, and shared interests in landscape and memory, the music that has been issued through Clay Pipe is less obviously dominated by the radiophonics of the 1960s and 1970s than Ghost Box, taking in a wider range of influences from modern classical, ambient, jazz, post-rock and minimalism, with strong nods towards pastoral and folk traditions. Highlighting the difference between Cate Brooks' releases on Clay Pipe and those as The Advisory Circle, Ian Preece hence describes the label as Ghost Box's 'semi-bucolic cousin': more innocent, less knowing and arguably more well-mannered.[1] Certainly, while many Ghost Box releases are eclectic and typified by sometimes jarring juxtapositions of style, Clay Pipe is, in comparison, easier listening. Most releases are instrumental, eschewing obvious samples in favour of subtle field recordings which provide texture and tone. Many are haunting, somnolent and meditative. The music always seems beautifully sequenced, designed to evoke the atmosphere of particular times and places rather than generating a disturbing sense of the uncanny. As Frances Castle tells me:

> I used to say that I put out 'atmospheric music with a strong theme or sense of place'. But in truth I've always found it hard to define exactly where the music I put out fits, and what links it all together . . . I'm not tied to any genre or style . . . I mainly put out instrumental music, something you can enjoy while in the landscape itself – you can listen while walking or driving, or on a train journey. It can act as a soundtrack to your surroundings in a way I don't think any other art form can.

While not all of Clay Pipe Music's releases concern landscape, the summoning of place and atmosphere is a common concern, one arguably tracing its roots to the first label release in 2010, The Hardy Tree's *The Fields Lie Sleeping Underneath*.

The Hardy Tree is Frances Castle's own recording project, and her first since Transistor Six, who released a series of quirky electro-pop singles in the late 1990s, including the intriguingly titled *Post Office Tower EP*. In between, she worked in the computer game industry, and then turned to commercial illustration, including book jackets and graphic short stories. As she recalls, there was no initial plan for Clay Pipe beyond the fact she saw the issuing of handprinted CDs in limited runs as a way to develop her art portfolio:

> I'm not sure exactly what my plans for Clay Pipe were in October 2010 when I finished recording The Hardy Tree's *The Fields Lie Sleeping Underneath* and decided to self-release it on CD. I handprinted the cover in blue, using a miniature Japanese screen printing machine, and I catalogued it Clay Pipe 001, so perhaps I was more optimistic about starting a label than I remember? It got picked up by Jarvis Cocker and played on his 'Sunday Service' show [*on BBC Radio Six*], the first pressing of 100 sold out, so I quickly pressed up 100 more, this time in yellow ochre. A precedent was set.

The Hardy Tree was named after the ash tree in St Pancras churchyard associated with the novelist Thomas Hardy. Working for a local London architect in the 1860s, Hardy was reputedly charged with relocating the graves of those displaced by the construction of the new Midland railway and decided to stack their gravestones in the centre of the churchyard, their bodies below in a collective grave. Over time, the growth of the ash tree between these graves became a symbol of life after death, its association with Hardy possibly apocryphal but celebrated nonetheless. When the tree fell in a 2022 storm, its loss was much felt by Hardy enthusiasts; the romantic myths around it kept alive by the planting of a new tree a few yards away, a symbol of the persistence of the past in the present. The name of the record label similarly alludes to this tenacity, referring to the fragile clay pipes often found preserved in the mud of the Thames embankment, collected by mud-larkers including Frances and her partner.

Such allusions to landscape history and the way the sublime can be encountered among urban ugliness informed The Hardy Tree's debut LP (Figure 3.1). *The Fields Lie Sleeping Underneath* (2010) provides a romantic exploration of forgotten London recorded in the attic room of Frances Castle's house. As Castle explains, this concerns her local neighbourhood of Finsbury Park, a Victorian suburb. Digging into local history, she began 'seeing things in the layers in the past', the record developing as a soundtrack to the past beneath the pavements. Some songs on it are named after former residents:

Figure 3.1 *The Fields Lie Sleeping Beneath* by The Hardy Tree. Courtesy: Frances Castle.

> I just looked on census returns and found the names of the people who lived here before. Walter Stokes was an astrologer. That was his hobby . . . He used to go around the UK with glass slides, giving lectures in halls and places about the stars. So that song ('Walter R. Stokes') was about imagining him during World War One when he was living in an attic flat, looking out of the window with a telescope looking at the stars . . . It's a secret history, isn't it? That something was left of all those people.[2]

The music is slow and subtly textured – simple piano and mellotron motifs, occasional violin, field recordings and airy woodwind – suggesting an autumnal amble through the leaf-strewn back lanes of suburban London, stumbling across all manner of arcane legends and stories. An urban pastoral, perhaps? The opening track 'Harmony Hall' features the sound of a squeaking gate, a possible homage to a sonic motif on Virginia Astley's (1982) influential pop pastoral *From Gardens Where We Feel Secure*, while on the title track an oboe lilts over a crackling, spoken lyric that ruminates on the past lives hidden beneath the cold concrete: the sleeping dead, the ghosts of the past and 'the furrows of the plough'.

Quirky, quaint and evocative, this album set the pattern for The Hardy Tree's subsequent releases, *Through the Passages of Time* (2016) and *Common Grounds* (2022). Both treat the urban landscape as a palimpsest, a layering of different stories and lives whose traces remain, sometimes physically but often merely psychically. The strong musical narrative on both is again reinforced by striking artwork betraying Frances Castle's background as a freelance children's illustrator. *Through the Passages of Time* invites us to listen carefully, creating a mood of quiet contemplation in which we can imagine the congregation that used to throng in a now-demolished church in Bermondsey ('St John Horsleydown'), or the Thames shipbreakers who worked at 'Baltic Wharf'. *Common Grounds* also hints at the importance of solitude in the city, based on a COVID-19 lockdown project where Frances Castle became an urban *flâneuse*, walking the streets and turning these daily excursions into demos she listened to during the following day's walk, allowing the music to shape her engagement with the city. Tellingly, the cover shows a view from a contemporary London window, replete with Sky satellite dishes and modern streetlamps, but the skies are dark with the shadows of the Second World War German bombers phalanxed in ominous formation.

Though The Hardy Tree's albums peel back the layers of a real place – London – another of Frances Castle's projects, *Stagdale*, is based on a Belbury-type composite. This is a village nestled in rolling countryside, very much the idealization of the English rural idyll: the settlement is overlooked by a stone circle with a white horse figure scoured from the chalk on the hillside, a Norman church below. As Castle suggests, 'It could be anywhere, really – any small, isolated village in the UK. I don't know if there are many places like that left any more . . . but there probably were in the 1970s'. *Stagdale* is a multi-part graphic novel, each accompanied by a flexi-disc of three instrumental tracks from The Hardy Tree (Figure 3.2). Together, these tell the story of Kathy, who moves to Stagdale following her parents' divorce in 1975. Her new cottage lacks TV reception, the squire is unhelpful and locals unfriendly. The village museum also reveals a history of superstition and ritual, much of it tied into a jewel supposedly stolen by a German emigree before the Second World War. Kathy then discovers a wartime diary – hidden behind the skirting board in her new home – which describes the escape of a boy, Max, who came to Stagdale as a refugee from Germany in the pre-war years. Here, the stories of two outsiders link, with Castle suggesting the children become 'tied together in a way that doesn't often happen in real life', albeit suggesting our own lives are always connected to those who preceded us. Musically, *Stagdale* juxtaposes 1970s rural England and Berlin in 1938: the 'Theme to Stagdale' is a gentle, piano and flute-driven tune invoking the English pastoral tradition, but this breaks down mid-song as the melody begins to reverse, and the closing seconds also hint

Figure 3.2 *Stagdale*. Courtesy: Frances Castle.

at something more sinister, with 'The Museum' cranking up the mystery with fluttering, spectral melodies.

Castle suggests the soundtrack to *Stagdale* is influenced by some of the unsettling 1970s children's programmes frequently cited as inspiring hauntological music, *The Changes* and *Children of the Stones* in particular. In *Stagdale,* the psychic bridge between two eras and the suggestion of temporal instability also fits with the hauntological critique of progress, as well as a fascination with the stains of history in the contemporary landscape. Here, *Stagdale* draws on Castle's childhood experiences, which centred around her family home in West Kew but also involved summer trips out of London. As she recalls,

> I remember visiting my grandparents in Sussex, and behind their house was a field – we used to walk down there a lot. That had a huge bomb crater, and I remember my Dad telling me that the planes would fly past to bomb London and then fly back over the English Channel . . . and they'd just let their bombs drop in the fields.[3]

Castle's other grandparents lived in Cookham, Berkshire, the Thames-side home of the artist Stanley Spencer, who referred to it as a 'village in Heaven'. Castle's maternal grandfather was also an artist: Frank Sherwin specialized in watercolours, selling many of his images to greeting card companies but remains well known for the posters he completed for British Railways in the 1950s (Figure 3.3). These posters were regarded by the railway companies commissioning them as fine art, being exhibited in gallery spaces as well as at local stations. These stations were, in effect, 'outside galleries' offering a Modern interpretation of an England that was located not just in the Southern lowlands but the rugged beauty of the Yorkshire Dales, the moors of Devon and the dark peaks of the Pennines.[4] Though Frances suggests she had only a dim awareness of these posters when she was a child, she now recognizes the similarities between these and some of the art she has produced for Clay Pipe.

Castle's art for Clay Pipe has a near-childlike naivety, portraying England as ostensibly bucolic and settled, but always with darker hints of change and loss, mirroring the work of Paul Nash or John Piper, inter-war artists who portrayed the English landscape as strangely devoid of human figures, enigmatic and uncertain. This ambivalence reflects the somewhat 'Janus-faced' response to Modernity that was prominent in the period, with critical commentators both mourning the impact of modern technology on the English landscape but

Figure 3.3 Frances Castle, Clay Pipe Records, pictured in front of her grandfather Frank Sherwin's 1950s posters for British Railways. Courtesy: Frances Castle.

also celebrating the fact it offered unfettered access to traditional agricultural landscapes and wild spaces otherwise untouched by 'progress'. Christopher Neve's classic book on twentieth-century landscape art, *Unquiet Landscape* (1990), suggests the work of Piper and Nash is Modern in the old-fashioned sense that the art looks newer and more original than much of the art being produced today, its abstraction of form imbuing it with near-surrealist qualities that hint at the ambivalence of Modernity. As Castle notes:

> I looked at a lot of early 20th Century British artists and printmakers – Eric Ravilious, John Minton, John Piper – and the images seemed to sit well and reflect the music I was listening to – mainly folk and experimental electronics. They were forgotten for a long time . . . Perhaps it's something we only see now that that makes them relevant. It's very hard to say . . . There's something though, a darkness there? There's something else. I'm not interested in just drawing a nice picture of a landscape.

For Castle, the cover art is crucial in putting the listener somewhere 'before they put the record on', setting up 'a space in their imagination for the music to play out'. The development of a particular Modern style clearly imbues Clay Pipe with a strong label identity. Inspired by other labels with a single-minded design aesthetic, particularly Glen Johnson's Second Language records, Castle's art has made Clay Pipe eminently collectable: the label is strongly oriented towards the selling of music as an artefact, with CDs, tapes and vinyl releases all hand-numbered, typically limited editions of 300–700. These often sell out in days, some scalped by eBay traders hoping to cash-in on the label's 'boutique' reputation.

Romantic moderns

While Frances Castle suggests she began with no clear vision for Clay Pipe, The Hardy Tree's interest in soundtracking forgotten landscapes has served as a model for many other releases on Clay Pipe Records. Some are re-releases of previously issued albums displaying a similar interest in the biography of place, such as Plinth's *Music for Smalls Lighthouse* (2013) by Michael Tanner, and the 2012 album *Tyneham House*, originally released by an anonymous group of musicians on Second Language records. Tyneham was a village in Dorset taken over by the military in 1943, and, despite promises to the contrary, the displaced villagers were never permitted to return. Now an abandoned village where 'time stands still', Patrick Wright's (1995) book *The Village That Died for England* suggests it has been a leitmotif in debates about national identity ever since. As he shows, for some its abandonment reveals the sacrifice rural folk made in the national interest, while for others, it symbolized the post-war encroachment of distinctly unpatriotic values on the English countryside.[5] The album *Tyneham House* gently prods at the ghosts of the village, the pastoralism of the inter-war period providing the basis of its musical vignettes of the times before the tanks came. The album celebrates the ordinary lives that villagers led before they were displaced from the land, 'throwing rocks into the big pond and grazing sheep in the rectory garden'.[6] The album is hauntological in the sense that it works through themes of ruination, but the musical inspiration is from the inter-war rather than the 1960s electronica of Radiophonic Workshop: the album is more Delius than Delian, wistful instrumental tracks consisting of gently strummed guitars, interwoven with flute and accordion.

Also on Clay Pipe, Darren Hayman's (2012) *Lido* is a similarly thoughtful exploration of the rise and fall of Modernity. Formerly of indie-pop band Hefner, a favourite of the late John Peel, Hayman's eclectic solo career has seen him write paeans to his native Essex (*Pram Town*, 2009; *Essex Arms*, 2010; *The Violence*,

2012), faded British coastal resorts (*Eastbourne Lights* EP, 2007) and those few English villages whose population was not diminished by military involvement in the First World War (*Thankful Villages* Volume One – Three, 2016–18). *Lido* is, in contrast to most of his output, entirely instrumental, gently strummed ukulele and guitar, with occasional violin and accordion, layered over field recordings made at outdoor swimming pools (e.g. Brockwell Park, Saltdean) or the sites where these used to be (e.g. Purley Way, Brentwood). The album begins with a babble of children's voices at London Fields, an optimistic and upbeat tune which celebrates the 2006 reopening of the 1932 Hackney Lido that was closed following cuts to the GLC in the 1980s but is now London's only heated outdoor Olympic-sized swimming pool. Hayman's music, coupled with Castle's stylized cover image of 1930s bathers luxuriating in a streamlined setting, celebrates the optimism of the inter-war Modernity that regarded civic improvement as a must, and invested heavily in facilities designed to promote the health and efficiency of the citizenry. Those tracks that reference long-disappeared lidos are hence infused with regret and melancholy: 'Brentwood', for example, is a woozy acoustic number recalling hazy days of summer long gone, the pool having been replaced by a Sainsbury's superstore in the late 1970s.

These themes of thwarted municipal progress are also highlighted on David Boulter's third solo LP, and his second for Clay Pipe, *St Ann's* (2024). Boulter, who now lives in Prague, grew up in Nottingham and was a founding member of The Tindersticks in the 1990s. A multi-instrumentalist best known as a keyboardist, Boulter has numerous soundtrack collaborations under his belt and a solo career that often explores the entwining of biography and place. *Lover's Walk* (2021, on Spoken Word Records), for example, was a deeply personal album about Nottingham, his return to the city to visit a dying former girlfriend the springboard for a pensive spoken word recording noting nothing has changed in the city since his youth, but that everything is different. His subdued chamber music evokes walks along the River Trent embankment, eloping in the Derbyshire countryside, and guilty pints in the station bar.

St. Ann's reflects on the redevelopment of the inner-city estate where Boulter grew up, his family displaced – alongside 10,000 others – when the estate was demolished and redeveloped on Modern lines when he was aged six. The record is deeply nostalgic about the streets he grew up in: the LP begins with the sound of children incanting the nursery rhyme 'Here we go looby loo' on 'Plantagent Street in the Morning', over a gently looping guitar melody. But Boulter doesn't romanticize the past, acknowledging the optimism felt about the redevelopment, which promised better living conditions and the eradication of poverty, but which quickly dissipated. A 1968 study had found widespread material deprivation in St Ann's, with only 9 per cent of homes having inside toilets and half still boiling up hot water on the hob for washing or bathing.[7] The new homes improved this situation, bringing Modern facilities to their inhabitants, but, some forty years later, sociologist Lisa McKenzie notes the hope that things would get better has vanished, the continuing stigmatization of the estate suggesting technological fixes were not sufficient to create a sense of belonging:

> If you move into the estate you realize that the old St Ann's with its narrow streets, haphazard layout and confused, illogical planning has been replaced with exactly the same only in grey concrete. The flats and maisonettes are suspended on footbridges on several levels and reach each other through underpasses and subways, which are very often dark and covered in rubbish and debris. Between each row of houses there are three-feet wide paths running down the back of the small square back yards; local people call them 'rat-runs'.[8]

Returning in 2022 when his mother could no longer remain in her home, Boulter similarly found an estate in post-COVID-19 decline, local stores and business boarded up. But while the corner shops and pubs are gone, he still senses some of the hopefulness he recalls from his youth: the music never falls into sonic clichés about the loss of the past, but is a meditative, calm and emotive exploration of a place once called home.

Twenty-first century reflections on the vanishing landscape of mid-century Modernism are also apparent on the Clay Pipe releases *Green Line* (2017) and *Adlestrop* (2020) by Gilroy Mere. This is one of the musical aliases of Oliver Cherer, who has also recorded as Dollboy and Australian Testing Labs, as well as playing bass in the Miki Berenyi Trio. In many ways, Clay Pipe seems the perfect fit for Cherer's conceptually focused soundscape music, with Frances admitting 'Gilroy Mere was pretty much invented for Clay Pipe'.[9] Like Castle, Cherer professes an enthusiasm for exploring the past in the present:

> Layers of fading history fascinate me. I love to discover prongs of the past poking through the fabric of now, unseen or adapted into invisibility. This stuff is everywhere. It might be of the past but so much is still here that it's part of the present too. For example, many of the kerbstones around my neighbourhood retain the marks of Victorian bath chair hire points. They're small lead-filled holes that used to hold metal posts. They're a reminder that the kerbs haven't changed in 150 years. I love that they're unseen and integral.

The first of Cherer's LPs for Clay Pipe – *The Green Line* – is an instrumental album about the buses linking London to the Home Counties between the 1930s to the 1980s, resplendent in their bright green livery. Cherer grew up in Croydon, South London, watching these buses ferrying Londoners out to Kent, Sussex and Surrey, placing him on an imaginative border between town and country and sending his mind spinning about the summertime pleasures to be found beyond the 1930s suburbs. John Grindrod, author of a celebration of British Modernity – *Concretopia* (2018) – found much to admire about the LP:

> This is a gem of an album: melodic, warm, multi-layered, with a driven narrative and a message about what we once made possible. In writing music about such an overlooked and forgotten piece of our cultural history, Cherer has done something quietly remarkable. He has resurrected the spirit of the Green Line and reminded us of extraordinary moments lost; everyday moments seldom captured in music, writing or film. You listen to it at home and it takes you on a journey. You listen to it outside and it heightens those moments. It's a bus ride you want to keep taking.[10]

Continuing, Grindrod waxes lyrical about the opening track 'Dunroamin', suggesting the song's ticking percussion 'perfectly orchestrates the rattling rivets and loose prefabricated panels of an ageing bus', 'evoking the anticipation and excitement at the start of a trip'. As he notes, the album is akin to Kraftwerk's *Autobahn*, albeit the record praises the arcadian pleasures of English country roads in the inter-war years rather than the high-tech fantasies associated with German motorways in the 1970s.

Quite why Cherer would want to make a carefully conceptualized record based on English buses is open to question, not least because nostalgia for buses, and bus-spotting in particular, is often maligned. Yet Cherer dismisses the accusation he's backward-looking:

> I really don't like the idea of being a nostalgist. I don't think 'the old days were better' and I don't live in the past. What I *am* interested in is the way our heritage helps to define or form who and where we are now. Old railway stations remain like scars that tell of past lives and landscapes and histories . . . On *Green Line*, I'm really dealing with memories rather than artefacts – the fact that it's a living memory is interesting to me – it informs who I am and hopefully others who remember too.[11]

The implication is that the *journeying* is more important for Cherer than the means of travel per se: the interest here is in social mobility set in motion through affordable public transport, and the connections made between places that otherwise existed in different social worlds. The track 'Hop Pickers', for example, alludes to the annual out-migration of Eastenders to Kent where they were offered basic food and lodging in return for a long summer 'holiday' spent picking in the orchards and fields; 'Ditchling Beacon' recalls day-tippers' journeys to the highest point in East Sussex, a site much beloved of those in the inter-war arts and craft movement; and 'Cuckoo Waltz' suggests a trip to a castle, possibly Windsor, soundtracked by the music of the 1950s old-time dance revival. *The Green Line* is then partly a celebration of the inter-war 'motoring pastoral' described by David Matless, a restless search for the beauty of the past in which motorized transport promised urbanites easy access to landscapes conforming to pictorial conventions of rural Englishness.[12]

On *The Green Line,* field recordings are used sparingly but effectively to ensure the experience of journeying is conveyed, with Cherer telling me the field recordings were added after the record was written and recorded. As he recalls, 'I lucked upon an old 60s sound effects EP in an auction just as I was finishing the recording which comprised a series of recordings of a Routemaster bus and it was too good an opportunity to pass up'. But elsewhere Cherer uses field recordings more as inspiration, with these encouraging forms of *active* listening:

> I first recorded environments on a portable minidisc recorder with a Sony stereo mic that seemed to 'swallow' sound and, on playback, revealed noises that I hadn't been aware of in the moment. I came to realised that, in a process similar to what happens with photography, the act of recording environmental sound makes me concentrate on those environments which open up to me as revelations. As a teacher, I've used this technique on students allowing them to break the sound of their surroundings into smaller and smaller components: one sound – for example, the breakers crashing on the beach – can be broken into three, four or five sounds and they gradually became more aware of the sound around them.

This modus operandi helped him produce his second Clay Pipe LP *Adlestrop* (2020), named after a now-absent railway station on the Gloucestershire-Oxfordshire border, itself the subject of Edward Thomas's 1914 lyrical poem celebrating the quiet peacefulness of England before the outbreak of the First World War. For *Adlestrop* Cherer visited far-flung former railway stations: Aldeborough in Suffolk, the delightfully named Star Crossing on the Mold &

Denbigh railway, Torver in Cumbria, Christ's Hospital in Sussex among them. As Cherer explains:

> I'd always loved the poem Adlestrop but it was seeing a copy of the 1962 Beeching report and its list of proposed station closures that was the spark. So many of the place names were beautiful, poetic and evocative of a faded Albion. I made field recordings in each of the stations on the album, or what was left of them. I then played these recordings while improvising in my studio, slowly building up a collection of pieces that I felt reflected the place and origin of the recording. Some things probably go unnoticed; for instance, the sound of traffic going over a join in a flyover. If it's heard at all it probably sounds like a quiet kick drum. I think a good deal of what goes into the mix on these things is inevitably overlooked ultimately, but it's about method.

This method helped Cherer to capture the ghosts of the stations and their enduring influence on the present. On an accompanying flexi EP, *Over the Tracks* (2020), one track alludes to his daily experiences of living in Hastings and passing the buddleia-covered platform of what was St Leonard's Marina station, now in the shadow of a retail park. As he describes, 'Swallows' attempts to evoke 'the stillness of that platform between trains in the summer, when the buddleia is a-buzz with birds, butterflies and bees. I always look out for the platform as I pass, knowing that it goes almost completely unnoticed by everyone else.'[13]

Adelstrop benefits from Castle's inventive sleeve illustrations, as well as a paper cut-out 'make your own station' kit in each copy (designed by Gary Willis). The release was also re-pressed in several different covers, the livery of the trains on the jacket matching the colour of the limited-edition vinyl inside, crimson, brown or blue to match the colour of British Rail's post-war carriages, Pullman carriages and those of Southern Railways respectively (Figure 3.4). Pictured inside the carriage is a solo traveller, travelling third class but dressed in a tie, tank-top and jacket: the image readily brings to mind the 1939 painting by Eric Ravilious, 'Train Carriage', a classic image of the inter-war landscape in which the Modern and ancient are bought into juxtaposition, the view from the train window towards the South Downs leading the eye to the chalk figure of the Westbury Horse (Figure 3.5). The chalk downland of South England was the landscape most inspiring for Ravilious, his work emerging in a period of enthusiastic topographic travel fuelled by both Shell guidebooks and British Railway posters, hillside chalk figures providing succor for his musings on the 'irrational, numinous forces of the English past'.[14]

Figure 3.4 *Adelstrop* front and reverse cover. Courtesy: Frances Castle.

Figure 3.5 Eric Ravilious 'Train Landscape'. Courtesy: Aberdeen Art Gallery / Creative Commons CC0 1.0.

Ravilious' imagery encapsulates the ambivalence of the inter-war period in its admiration and condemnation of Modern transport, both in awe of the Westbury horse and its mysterious origins but also suggesting its power had been diminished by the arrival of the train, which rendered it the tamed subject of the tourist gaze.[15]

But in Castle's illustration, the seated figure is not Ravilious, but Edward Thomas, the author of the poem 'Adlestrop'. Thomas was a perennial essayist who frequently travelled out from his Hampshire home, his books – notably *The South Country* (1909) – aiming to document the landscape of 'the North Downs and the South Downs, the Icknield Way and the Pilgrims' Way . . . a land of hops, fruit, corn, high pasture, meadow, woodland, heath and shore'.[16] He is remembered today as the soldier poet who registered the ambivalence of Modernity, gauging its erosive effect on rural communities but celebrating its ability to transport him into 'deep' England. His summation of the train as both a symbol of progress and a wrecker of rural life was captured in 'Adlestrop', written when his train made an unplanned stop at the station, forcing him to observe the stillness of nature in an otherwise unremarkable but quaintly named English village. A recital of the poem, with languid ambient backing, is a centrepiece of Gilroy Mere's album:

> Yes. I remember Adlestrop –
> The name, because one afternoon
> Of heat the express-train drew up there
> Unwontedly. It was late June.
>
> The steam hissed. Someone cleared his throat.
> No one left and no one came
> On the bare platform. What I saw
> Was Adlestrop – only the name.
> . . .
> And for that minute a blackbird sang
> Close by, and round him, mistier,
> Farther and farther, all the birds
> Of Oxfordshire and Gloucestershire.

Here Thomas captures 'a disappearing, encroached-upon English countryside, preserved for eternity in dark, eerie, haunted, uncanny verse, beset with loss', albeit the poem also summarizes the mundane pleasures of travel.[17]

Thomas died in the Battle of Arras, 1917, but the poem is now inscribed on a plaque beneath Adlestrop station sign – albeit the station was closed in 1966 and the sign is attached to a bus stop. Cherer is too young to have travelled there by train, and was born long after the Beeching Cuts, but the poem still clearly resonates with him, capturing the way that an embodied encounter with landscape can trigger all number of memories and feelings, its psychic charge oddly inexplicable:

> I have it in my head that it represents a moment in time immediately before the chaos, horror and noise of the First World War and it perfectly captures that. That moment where the great steaming beast of the train is frozen and 'all the birds of Oxfordshire and Gloucestershire' combine in a heat haze with someone unseen clearing their throat. It is very poignant, and I've occasionally had those moments where I was suddenly very aware of everything around me and wondered if I would remember it forever.[18]

Like *The Green Line*, *Adlestrop* was well-received, reviewers commenting on its 'ochre-toned arcs of electronic breaths and the gentle timbre of English oak . . . the divine rustle of bees and the paradisiacal birdsong of Oxfordshire and Gloucestershire'.[19] Clearly, it's more than just an evocation of lost train stations.

Oliver Cherer's work as Gilroy Mere concerns the persistence of a past version of Englishness in the present, and the way that current concerns and crises – including those relating to railway nationalization, High Speed Two and Levelling Up – are layered on past national imaginaries. He tells me:

> I don't mind the term 'hauntological' if that label denotes an approach which I have adopted then it seems fair enough. But I remember discovering the word 'palimpsest' and a light bulb going on. I think that while I am interested in the past, it's really more about how some of that past is here in the present. I live a modern life in an old house. My shop is in a building designed and built in the mid-thirties [*the De la Warr Pavilion in Bexhill*]. The past is not better, it's still here. Sometimes it pokes through more recent layers. Sometimes it's still in use, unchanged and sometimes its traces are all but obliterated. This is what I find fascinating and where my own version of 'hauntology' lies.

So while Cherer's releases are personal, biographic reflections on place and not overtly informed by hauntological theory, the combination of the sleeve art and melancholic music, often borrowing motifs from the inter-war and immediate post-war period, hints at an interest in English landscape's endless capacity to conceal the social relations that produced it.[20]

In the country

Matthew Davies, recording under the name Vic Mars, has also released a trio of albums on Clay Pipe Music with more than a passing interest in the landscape. These albums – *The Land and the Garden* (2015), *Inner Roads and Outer Paths*

(2019) and *The Beacons* (2023) – centre on revisitation of the locations looming largest in his personal mental map. Davies' Herefordshire childhood is certainly recalled warmly: he remembers being able to see an Iron Age hillfort from his childhood bedroom, recollecting school holiday excursions into the country lanes and deep woods surrounding his home:

> The countryside is beautiful, and easily accessible, just a five or ten-minute cycle ride from where my parents live. The River Wye runs through the city [*of Hereford*] too, and on a good day it's quite picturesque . . . Growing up, there were always abandoned houses, ruined barns, bunkers, woods for camping and weird local legends. The munitions factory was the big one for us. A huge hangar, blast walls, bunkers, all sorts of stuff. And overgrown paths, so it was easy to get a bit lost in there . . . it covered a big area. There's also a church nearby, and we were told they took the hands off the clock to stop the Devil visiting at midnight. Weird stuff.[21]

Davies ended up teaching in Japan, releasing an electroacoustic LP (*Kanrasha*) for Japanese label Symbolic Interaction in 2012, but on one visit back to the UK became fascinated by some reprints of vintage British Railway posters hanging in his brother's house. This led him to a musical exploration of the landscape of this corner of England, immersing himself in the music of classical pastoral composers including Holst and Vaughan Williams, as well as the music of Donovan, Nick Drake, Can, Julian Cope, Moondog, Wim Mertens and Bert Jansch. Alongside this was the palpable influence of children's BBC soundtracks such as those by Freddie Phillips for *Chigley* and *Trumpton*, and Vernon Elliott's *Ivor the Engine* and *Noggin the Nog*.

These disparate influences were evident on Vic Mars' 2015 album *The Land and the Garden*. With titles including 'Villages, Hamlets and Fetes', 'Butterflies, Bees and Other insects' and 'Hedgerows and Conservation', this combined gentle glockenspiel, recorder and synthesizer to create a lilting sense of a timeless rural: the accompanying video for 'Road through the Village' accentuated this, featuring footage culled from 1950s transport films, a brass band playing at a village fete, cream teas served on gingham tablecloths, brown trout flitting around a chalk-bottomed stream. This album was compared by some reviewers to Virginia Astley's (1982) *From Gardens Where We Feel Secure*, Mars admitting 'the titles of the tracks were quite descriptive and to the point, it was in general a kind of hazy, catch-all view of Albion, fields and trees, farms and hedgerows unblighted by new roads and construction'.

This suggests *The Land and the Garden* was a borne of a nostalgic desire to return to an Englishness located in a particular time and place. Vic tells me:

> It all began from homesickness while living in Japan. Appreciating more when away how beautiful the countryside is and realising how much I missed it . . . I didn't go out of my way to emulate British composers and use similar chord progressions or modes, but it's the instrumentation, flutes, clarinets etc, mellotrons which makes it feel British . . . Going back to people that have been for walks and listened to my music, hopefully the music connects them to their surroundings and will remind them when they hear it again elsewhere. Music has the ability I think, when you hear it first, whatever you are doing, wherever you are, it will bring back that memory later on, so it's great if people hear my music on a journey somewhere away from the city and suburbs.

Mars juxtaposes urban and rural, the former regarded as a threat to rural ways of life that seem more authentic and appealing than city living. The countryside is here portrayed through a musical idiom that uses classical pastoral imagery to emphasize both continuity and change, but in so doing it potentially falls into the trap of becoming sentimental and nostalgic about an England that has disappeared.

This preoccupation with the disappearance of the remembered landscapes of youth situates this album in the overlap between pastoralism and hauntology. Tellingly, the cover of the sequel, 2019's *Inner Roads, Outer Paths* is not a photo but an image somewhat typical of children's books from times past: simple lines, bold colours, almost bucolic in its feel, somewhere between Alfred Bestall's magical landscapes in the *Rupert* annuals of the 1960s and the British Railways posters that encouraged travellers to visit the far-flung corners of the country (Figure 3.6). Castle's image shows a neat, low-rise post-war estate of uniform bungalows of the type that surrounded many English towns and cities in the 1960s, complete with a dismal municipal playground in the foreground, rolling green hills in the background. The suggestion here is that the country has been lost to the town, but the details speak of other possible narratives, the cement mixer at side of road hinting at ongoing building while on the hillside there is a row of similar housing next to the ruins of stately home, possibly the one referred to on the track 'The Last Days of the Great House'. The back cover changes perspective so we are looking down on new, modern homes through the broken window of the abandoned house, posing the question of who, or what, is watching? Perhaps we are being encouraged to imagine the young Vic Mars

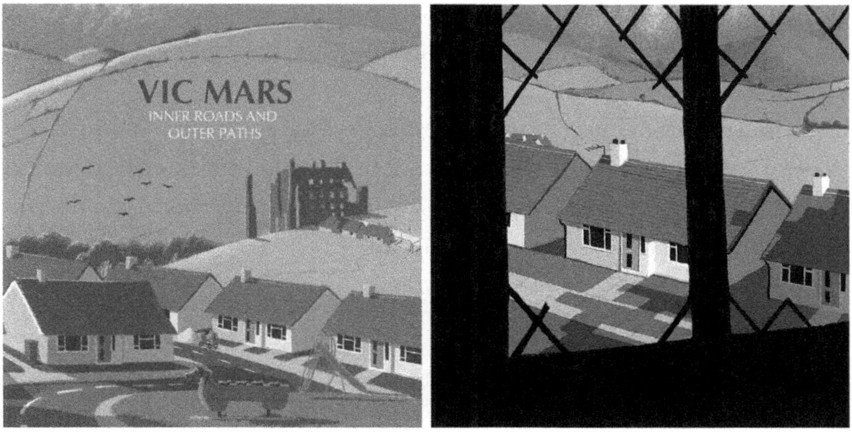

Figure 3.6 *Inner Roads, Outer Paths.* Courtesy of Frances Castle.

having broken into this deserted hall, surveying the new housing obliterating the familiar landscape below.

On *Inner Roads and Outer Paths*, the distinction between town and country, and past and present, appears less clearcut than on its predecessor: on a return visit to Herefordshire, Vic had found a new road cutting through the fields he had played in as a child, and realized some of the older pathways had become lost or overgrown. The album hence works with ideas of disappearance and the impossibility of return, the landscape a composite of things that remain the same and those that have drastically changed.[22] Here Mars was partly inspired by Alfred Watkins, the Hereford-born polymath and photographer who got to know the county intimately through his travels as a representative of his family's brewing and bread-making businesses. Watkins' interest in photography led him to develop an eye for the antiquarian in the landscape, and he became the president of Herefordshire's natural history and archaeology service in 1918, writing a series of articles on dovecotes, wayside crosses and old chapels. Later, in a 'rush of revelations', he began to see connections in the landscape between certain features, suggesting their siting was not piecemeal but engineered and deliberate. His book *The Long Straight Track* (1925), illustrated with photos of his native county, offered perhaps the clearest articulation of the idea that certain topographical features – castles, monuments, ponds, churches, burial mounds, hedgerows – could be mapped through a series of alignments or 'ley lines'.[23] Watkins' ideas were often derided by the archaeology community, but triggered an unlikely 1920s craze as devotees scoured the countryside searching for alignments.[24] In the 1960s these ideas were rediscovered by a new generation, sometimes imbued with mystical overtones as writers hypothesized connections between ley lines, dowsing lines and the occult.

The music on the album summons journeys made along the trails, roads and hollow ways that lead out from Hereford into the countryside, excursions into a recent past that resonates with more ancient times. For Vic Mars, Watkins' sensibility to the landscape was inspirational:

> I recently re-read *The Old Straight Track*. Watkins was an interesting character, not only for his studies . . . but he also invented a light meter for cameras. Seeing his photographs and reading of places I vaguely knew, or had visited, gave me a feeling of history and how things seemed more peaceful and slower paced in his day.[25]

Vic Mars' album displays an evident interest in ruination, emphasizing the way the landscapes of the past remain in the present despite incessant development and urban expansion. The album harks back to a period in Vic's youth spent 'exploring the abandoned houses and factories on the fringes of his home town; the in-between places where nature either takes back, or loses its grip'.[26] Tracks such as 'A nest in the Warehouse Roof', 'Broken Spires and Ruined Arches' and 'End of the Branch Line', confirm this interest in decay and obsolescence. Gentle glockenspiel, recorder and guitar, sequenced through 1980s tracker software, combine to provide a sometimes mournful and contemplative elegy to the relict landscapes of the urban fringes: the ruins of the Mother's Pride Factory built in 1958 on Three Elms Road but demolished a quarter of a century later to make way for a DIY store; the former route of the Hereford to Ross-on-Wye railway that fell to the Beeching Cuts; and the remains of the Rotherwas munitions factory, the site of a deadly 1944 explosion.

Like the work of Oliver Cherer as Gilroy Mere, and Frances Castle's own Hardy Tree project, Vic Mars' work is distinctly English in its subject matter, mode of production and conceptualization. A key question sometimes levelled at Clay Pipe Music is whether this represents a form of nostalgia for a disappearing England, something that could be seized upon by those who harken back to a dream of the England that existed before a time of mass immigration in the 1960s. Castle recognizes this danger but dismisses it, placing Vic Mars work in the context of a more radical tradition that reclaims the nation in the name of the outsiders, the working-class and the weird. As she explains, 'There's a thread of slightly weird psychedelic English pastoralism that goes back to the 60s and beyond, music that was based in the hills of Herefordshire where Vic grew up . . . It hadn't really crossed my mind that Vic's music would be of much interest to those trying to stir up nationalism or patriotism'.[27] But when Frances Castle tells me Clay Pipe Music represents a 'form of escapism from the toxicity of things like Brexit, the political situation, and perhaps the over-amplification of modern life', this suggests she has given careful thought to the cultural politics of Clay Pipe's landscape vision and recognizes the potential for it to be interpreted as a retreat into older, regressive versions of Englishness. Oliver Cherer is equally mindful of the dangers of playing with the imagery of the English landscape:

> I think I probably am very British, English even. I'm not nationalistic or even very patriotic and have become increasingly uncomfortable with the idea of perpetuating the myth of a golden Albion. Is what we do (if we are psychogeographers or hauntologists) a form of eco-fascism? Is the Britain that my music evokes an imagined white one that a certain kind of nationalist yearns for? So much of what I've produced is about local history and therefore ideas and places that have disappeared. An album inspired by the Beeching cuts could be an analogy for the destruction of 'the old ways' or when Britain was 'great' While I can't see my music ever being used as a rallying cry for the Britain First crowd it does occur to me that perhaps I should be careful with this kind of nostalgia.

In the wake of the 2024 General Election, when the Reform Party won more of the popular vote than the Liberal Democrats, and Nigel Farage entered parliament for the first time, it remains important to air these concerns and acknowledge the possibility that even 'radical nostalgia' might be co-opted to exclusionary and racist ends. Clay Pipe's explorations of the lost landscapes of English Modernity are then celebrations of the social mobility ushered in by innovations in technology, building and government in the mid-twentieth century, but the way they play with the tensions of urban and rural implies an ambivalence that strikes at that heart of the failed Modern project. It's this ambivalence that is to the fore in other musical explorations of twentieth-century Modernism, not least those focusing on the failed urban utopias of the post-war years.

Further Reading

Harris, Alexandra (2023). *Romantic Moderns: English writers, artists and the imagination from Virginia Woolf to John Piper*. Thames & Hudson.

Matless, David (2016). *Landscape and Englishness: Second expanded edition*. Reaktion.

Neve, Christopher (1995). *Unquiet Landscape: Places and ideas in 20th-century British painting*. Thames & Hudson.

Owens, Susan (2020). *Spirit of Place: Artists, writers and the British landscape*. Thames & Hudson.

Wilkes, James (2014). *A Fractured Landscape of Modernity: Culture and conflict in the Isle of Purbeck*. Springer.

Wright, Patrick (1995). *The Village that Died for England*. Cape.

4

A Broken Concrete Utopia

Laura Oldfield Ford's *Savage Messiah* fanzine, published between 2005 and 2009, is often cited as a paradigmatic example of hauntology. A cut-and-paste montage of handwritten notes and sketches – part urban driftwork, part graphic novel – this is an account of how London's Olympic boroughs were 'hollowed out' in preparation for the sporting spectacle of 2012. *Savage Messiah* mapped out a landscape of derelict tower blocks awaiting demolition, their working-class tenants having been 'decanted' to estates in other boroughs, or even out of London.[1] In some cases, these tower blocks had been erected only a couple of decades earlier, once acclaimed Modernist estates that had promised every convenience but ended up stigmatized as housing for the poorest: sink estates for the underclass, in effect. For Oldfield Ford, these dystopian Modern landscapes, lying in ruin, were haunted by the ghosts of urban seditionaries and visionaries of yesteryear who had prospered in these affordable and inclusive urban environments:

> I was drifting through a London haunted by traces and remnants of rave, anarcho-punk scenes and hybrid subcultures at a time when all these incongruous urban regeneration schemes were happening. The idea that I was moving through a spectral city was really strong, as if everything prosaic and dull about the New Labour version of the city was being resisted by these ghosts of brutalist architecture, of '90s convoy culture, rave scenes, '80s political movements and a virulent black economy of scavengers, peddlers and shoplifters. I think *Savage Messiah* [*existed in*] the aftermath of an era where residues and traces of euphoric moments haunt a melancholy landscape.

Yet even as these landscapes disappeared beneath the corporate scenography of Olympic spectacle, she found hope in the ruins.

Savage Messiah is London recomposed, a psychogeographic perambulation that serves as an elegy for the East End and its once 'euphoric' Brutalist tower blocks. Tellingly, Mark Fisher provided the introduction to a 2011 Verso compilation of *Savage Messiah* zines, proclaiming Oldfield Ford's uncanny ability to see the contours of another world through the 'cracks and gaps' of an occupied London. Entitled 'Always Yearning for the Time that Just Alluded Us', Fisher's essay suggested *Savage Messiah* emerged at a time of 'bland delirium': namely, the belief that London's reputation as a creative hothouse could be financialized, its urban landscape transformed into a virtuous money-making machine where the working poor would become part of the creative class. Some hope. Instead, neoliberal urban policy turned London into a property hotspot, a store of value for overseas investors that priced out local populations, not least the fugitive artists who fuelled the diverse subcultures of the East End.

It's easy to see what attracted Fisher to *Savage Messiah*, its collage-like representations juxtaposing past and present: as Fisher argued, this emphasis on temporality is important in understanding the outcomes of what is euphemistically termed 'urban regeneration'. As Fisher recognized, 'the struggle over space is also a struggle over time and who controls it: resist neoliberal modernization and (so we are told) you consign yourself to the past'. Here, the derelict tower blocks became emblematic of the evisceration of the counter-public, with the transgression of the past transformed into a discourse adding value to the hipster-scripted East End 'renaissance':

> With their lovingly reproduced junk-strata, overgrowing vegetation and derelict spaces, *Savage Messiah*'s images offer a direct riposte to the slick digital images which the Olympic Delivery Authority has pasted up in the now heavily policed, restricted and surveilled Lea Valley. Blair's Cool Britannia provides the template for an anodyne vision of London designed by the 'creative industries'. Everything comes back as an advertising campaign. It isn't just that the alternatives are written over, or out: it is that they return as their own simulacra.[2]

The graffiti that constituted the vernacular language of the East End was a case in point. Situationist stencil art spelling out the provocation 'Gentrify This' was no longer a comment on the maelstrom of speculative urban development, but a de facto invitation to developers to exploit urban edginess, to find value among the ruins of Modernity.

Oldfield Ford's summoning of the ruins of Modernity is near-romantic, and neo-Ballardian. Her exploration of Modernity's ruins revels in the contradictions of the obsolete and overgrown, the remains of the planner's once-bold Brutalist dreams. Though not advertised as such, Pye Corner Audio's 2012 *Sleep Games*, on Ghost Box Records, could be regarded *Savage Messiah's* soundtrack (Figure 4.1).

Figure 4.1 Pye Corner Audio's *Sleep Games.* Design: Julian House, Courtesy: Ghost Box.

Eschewing the samples that characterize many Ghost Box releases, *Sleep Games* appears more in thrall to the electronic music of the rave generation: in place of the Radiophonic Workshop, *Sleep Games* references the techno of Burial, Orbital and Underworld, and offers a slurry, narcoleptic trawl through a sink estate on the edge of a post-industrial city. This estate is occupied by wraith-like figures, the ghosts of a thousand late-night raves. In 'Underneath the Dance Floor', persistent, almost-tribal drums are heard as if a party is ongoing in the flat upstairs, cold-wave synths drifting over the cacophony. CCTV systems appear broken, snowed with static. Sirens wail in the distance ('Black Mill Video Tape'). 'Nature Reclaims the Town', which closes the album, alludes to the common trope in science fiction of a city colonized by nature, the once-proud precincts of the Modern city now windswept and strewn with weeds.

Here, Pye Corner Audio's hauntology taps into a notable tradition of writing about declining cities, which, part inspired by Ballard, was thoroughly explored in the post-punk era, most obviously in the foreboding necromancy of Joy Division but also in the music of less celebrated acts. Josef K's solitary album, 1981's *The Only Fun In Town*, for example, was an appropriately Kafkaesque exegesis of urban ennui, tracks like 'Crazy to Exist', '16 Years' and 'Citizens' taut evocations of cities where 'memories are laid to waste'. But here the cities in question – Manchester and Edinburgh – might as well have been Berlin or Prague. The music exuded a gloomy European Modernism, with Ian Curtis and Paul Haig's existential angst evoking an era when cities boasted 'shoe shops and barbers and tobacconists, fancy confectioners, silk scarves and outfitters, department stores, restaurants deluxe, grand hotels', the type of urban fragments still visible in Britain in 1981 but which appeared increasingly phantasmic as cities disappeared under a sea of concrete.[3]

In contrast, *Sleep Games* dwells on the neoliberalism ushered in by Thatcherism, documenting how it ruptured the consensus of 'welfare state Modernism' to produce twenty-first-century wastelands known through the argot of 'sink estates' and 'no-go' housing: desolate precincts, abandoned shopping trolleys, pound shops, broken bottles. But the reference points are generic and hazy. The album is an exemplary exercise in conjuring mood and atmosphere, and is assuredly hauntological, but regards Modernism as virtuality. Modern planning appears as a failed utopian concept haunting present realities rather than something pointing the way to the future.

In contrast, some other musical projects have explored the ambitions of Modern planners at source, mapping out what *could have been*. Town & Country's *Plan for Plymouth*, for example, was released by Woodford Halse in 2022 (Figure 4.2). Housed in retro-styled *faux* cigarette packaging, a card inside profiles the 14-storey Civic Centre on Armada Way, completed in 1962 to replace the municipal offices destroyed in the Plymouth Blitz. Positioned on a grand North-South axis, the Civic Centre's opening marked the completion of Sir Patrick Abercrombie's bold plan for the city, co-authored with City Engineer James Paton-Watson to ensure 'the daily civic and business life of the city' would proceed smoothly into the future.

Like many of other plans forged in the wake of the Second World War and the emerging social consensus of the welfare state, the far-reaching scheme for Plymouth embodied a new form of social governmentality intended to reconcile the oppositions of town and country, tradition and Modernity, and economy and society, designed to assuage the class conflicts that had beset pre-war Britain. Here, a new notion of community came to the fore, based on the ideal

Figure 4.2 *A Plan for Plymouth.* Graphics: Nick Taylor, Photo: Terry Grundy.

of the neighbourhood unit, a village-like entity which would provide a balanced landscape for 'living, working and playing'.[4] On *A Plan for Plymouth* album, the upbeat Kraftwerk-like instrumental 'Ernesettle Bush' refers to one of the city's seven main post-war housing estates, each of 6,000–10,000 people, and each separated from the others by a narrow green belt. Developed adjacent to the Bush Radio factory, Ernesettle was built around a circular village green, its bustling shopping parade, adjacent health centre and school designed as a 'community hub'. This was thought by planners to be a socially efficient form of urban design, believed necessary to counter the social alienation characteristic of many pre-war cities.[5] As Abercrombie and Paton-Watson wrote, the creation of 'community' was intended to develop 'the spirit of companionship unknown to the youth of yesterday who vainly sought it in the car or the cinema', with the 'individuality of the citizen . . . encouraged and moulded into the community' through 'the right sort of facilities'. As they concluded, 'a plan which allows for a higher standard of living is well within our grasp, with its call for space and beauty rather than for mere economy'.[6]

A Plan for Plymouth's track listing namechecks many of the new landmarks of the post-war city, 'Inter-city house' (completed 1962) and 'Caradon Hill', the 700-foot radio transmitter that first brought ITV to Devon in 1961, among them. The latter is an appropriately upbeat electronic instrumental, but elsewhere there are hints of darkness. A video for the opening track 'Bretonside Rain' borrows 1962 black and white footage of a young girl walking slowly up the stairs into a deserted Civic Centre, a publicity film designed to introduce the new complex to the public.[7] The Centre is shining new, the girl standing in awe in front of impressive public art, dwarfed by the scale of the council's debating chamber. But the music here hints at faded grandeur and failed ambition: the Civic Centre was abandoned in 2015 and at the time of writing is being redeveloped by Urban Splash as 144 flats, having been sold by the Council for just £1.

Gateway to the future?

While *A Plan for Plymouth* was a one-off, a low-key cassette-only exploration of Modern town planning, Gordon Chapman-Fox (Figure 4.3), recording as Warrington-Runcorn New Town Development Plan (WRNTDP), has completed not just one but a whole series of albums about the dreams and failures of Modern planning, focusing on the twin New Towns of Runcorn and

Figure 4.3 Warrington-Runcorn New Town Development Plan. Photo: Meg Kerr.

Warrington. The former was one of the second generation of New Towns from the early 1960s, an overspill town for Liverpool, while the latter was part of the final wave of New Towns in the late 1960s, intended as a new growth pole for the North-West. Though both were originally designed to be balanced, free-standing communities, they merged for administrative purposes in 1981, with the Warrington and Runcorn Development Corporation finally dissolved in 1989.

Proclaimed as 'music for a broken concrete utopia', WRNTDP announced its arrival with the 2021 album *Interim Report, March 1979*. This was conceived as the soundtrack to a 1970s public information film assessing the progress made in creating the new communities, summoning the mood of the times and thoroughly grounded in the aesthetics of the 1970s. Part ambient Tangerine Dream ambient synth and part-slurry Radiophonics, the album begins with 'Gateway to The North', a track that imagines the New Towns programme as if it had been soundtracked by John Carpenter, all arpeggiated synths and woozy, tremulous bass. The track is presaged by the voice of planner Arthur Ling, the author of the 1967 Runcorn Masterplan. His voice, clipped and patrician, proclaims the vision behind the New Towns: 'It is an endeavour to provide a balanced town in which the motor car has certain privileges, public transport has certain privileges and people, perhaps most of all, have the privilege to walk about in safety'. Drenched in reverb, his words are sampled from a 1974 film by the Canadian Film Board on Basingstoke and Runcorn New Town, with Ling, a lifelong socialist and former Chief Architect of Coventry, among those interviewed about the early ambitions for the town. Very much the embodiment of the post-war planner, suited and moustached, in the film he stands in front of the town's master plan, with new 'neighbourhood units' subsuming the Cheshire countryside: Windmill Hill, Sandymoor, Castlefields, Beechwood.

Live, WRNTDP play in front of period footage of the New Towns. A promotional video for the WRNTDP track 'Gateway to the North', for example, uses clips showcasing the revolutionary segregation of vehicular traffic and pedestrians in Runcorn, intended to create an ordered, safe and commodious living environment. Much of the three-minute video, viewable on YouTube, focuses on the infamous Busway, the world's first 'Bus Rapid Transport' system, opened in 1971.[8] The footage shows the town as viewed from the drab 1970s beige-brown interior of the bus as it smoothly progresses on traffic-free roads through spacious modern housing estates criss-crossed by paths and 'pedways' following the rolling contours of the land. We then see clips of Shopping City, opened in 1972, a covered mall with connected multi-storey parking and bus station. It appears clean, streamlined and machinic, its futurist aesthetic at odds with the Austin Maxis and Hillman Imps leaving the car park. Without the final credits at the end imploring us to 'visit www.warringtonruncorn.com', the clip might easily be mistaken as from the 1970s, with the music seemingly of the time. Chapman-Fox recalls the recording process: 'OK, so what synthesisers existed in 1979?. I was . . . thinking if there was a guy in a recording studio in Warrington what would he have? He's got an eight-track at best and maybe two or three synths, there's probably a Hammond and an old mellotron that he got in 1967 gathering dust in the corner'.[9]

Musically, WRNTDP's use of analogue synth sounds from the 1970s takes the listener back to the time when these represented the musical cutting edge but were more likely to be soundtracking TV drama than heard in the music charts. Here, Chapman-Fox notes the influence of the Radiophonic Workshop:

> My music is very much inspired by my memories of the music from *Doctor Who* and other eerie programmes the Radiophonic Workshop soundtracked. I haven't made any diligent study or investigation into the exact equipment used, compositional techniques, or listened to hours of soundtrack albums. I just wanted to make music that captured that essence of that. Musical impressionism, if you will.

In doing so, he follows those Ghost Box Records artists whose citation of Radiophonic themes and styles aims to capture the atmosphere of the 1970s and 1980s: in the case of WRNTDP this associates this period with the beginning of the end of British Modern ambition.

The electronic experimentation of the 1970s and 1980s manifests in WRNTDP's music in different ways. On the debut album, ambient synth creates contemplative soundscapes which quietly hint at discontents to come, while on the follow-up, *People and Industry* (2021), the elegant melancholy of 'This Is the Age of the Train' speaks to the privatization of the railways and the debacle of High Speed Two which has left the North-West seemingly less connected to London than ever before. On the second LP, the music is fuller, and arguably more realized than on WRNTDP's debut, but the overall tone remains elegiac, hinting at themes of loss: 'Eye See Eye' is a quiet electronic throb overlain with delicate synth melodies, referencing the ICI chemical works that were for many years the most important employers in the area, while 'Managed Decline' growls ominously, a lugubrious tune that summons images of abandoned, rusting refineries. On the third WRNTDP LP, *Districts, Roads, Open Spaces* (2022), this electronic throb gives way to something more akin to Pye Corner Audio's hypnogogic *Sleep Games*, decayed pop with shades of Boards of Canada in the mix. Inspired by one of Chapman-Fox's visits to Warrington after pandemic lockdown, the music plays on themes of urban anomie and isolation. It also references the fact that the zoning of land use often led to women being isolated in the suburbs while men laboured elsewhere the city, resulting in the diagnosis of the 'New Town Blues' among young mothers.[10] The futures that seemed assured for young people also unravelled as the New Towns became stigmatized as failed social experiments: in the video for 'Old Hall' black and white footage of children playing beneath tower blocks is slowed down, reversed and remixed, as if time is out of joint, while the thirteen-minute opening track 'Golden Square', named after one of Warrington's 1970s shopping centres, is a slow-building atmospheric track that feels cold and empty. The once bustling precincts lie uninhabited and void, a few mini-marts and cash convertors hanging on for dear life.

As a playful, sometimes uncanny, citation of the past, *Warrington-Runcorn New Town Development Plan* is an exemplary exercise in hauntology, albeit Chapman-Fox describes this as a 'starting point rather than an exact, rigorous discipline'. Unlike much of the music in the hauntological tradition, WRNTDP eschews overt exploration of the supernatural or folk horror, instead soundtracking the mundane realities of the civic modernity, articulating how it sought to create a better society at the level of the everyday. In many ways, WRNTDP works in the gap between vision and reality. For example, Chapman-Fox celebrates the transport infrastructures that now appear anachronisms, remnants of a different, more optimistic, time: the aforementioned Runcorn Busway 'figure of eight' has become WRNTDP's de facto logo, appearing on t-shirts, postcards and vinyl slipmats, while also lending its name to a reverb-saturated track on WRNTDP's fourth LP, 2023's *The Nation's Most Central Location*:

> Lost futures and lost opportunities are at the heart of my music. 'Busway' is a bittersweet euphoric track that I've been finishing my live shows with, and it always makes me emotional. It's odd to write an emotional piece of music about a bus lane, but it all feeds into the central idea that these things were opportunities to improve our lives that were squandered.[11]

Chapman-Fox himself grew up in Chorley, on the edge of the proposed Central Lancashire New Town intended to merge Chorley with Preston and Leyland;

the New Town of Skelmersdale was just over the hill. He also regularly visited Preston, its futuristic bus station a monument to Modernity that ended up a shelter for the homeless, its rubber floors deteriorating thanks to its use as a de facto late-night urinal. These experiences seemingly inform Chapman-Fox's music, which comments both on the flawed realization of Modern ambition, and the gradual erosion of the communal values it sought to promote. In Chapman-Fox's words, 'When I was a child, the future seemed exciting. Now I'm an adult, the future seems terrifying. And I would like to go back to a time when the future seemed optimistic . . . I'm nostalgic for an optimistic future'.[12]

As critics have noted, WRNTDP provides a brilliantly conceived exploration of Modernism's legacies based on a pinpoint accurate reconstruction of aesthetic and sonic motifs from the 1970s and 1980s. The *Interim Report, March 1979* album sleeve, for example, mimics a late 1970s planning report, monochromatic photos beneath Letraset titles: for added verisimilitude, the sleeve is stamped 'Ex Libris Warrington Public library'. The idea that this is the soundtrack to a documentary film taking stock of the progress of the New Towns in 1979 is reflected in music situated on the cusp of euphoria and melancholia. Speaking to *The Planner*, Gordon Chapman-Fox notes that 1979 was when the New Town vision of balanced communities held together by communitarian idealism began to look badly out of step with the neoliberalism associated with Thatcherism. March 1979 witnessed the decisive vote of no confidence in Prime Minister Jim Callaghan following the 'winter of discontent': this ultimately opened the door for that year's General Election which saw the Conservatives sweep to power, resulting in the triumph of a model of individualism, as Chapman-Fox explains:

> That change from community to individuality is one of the reasons why, certainly in Runcorn's case, it didn't work out really as well as it could have done because people want their own cars. Runcorn was designed with its own separate bus lanes around the city and people instantly wanted to move away from commuting or travel by bus and wanting to have a car. The big concrete Brutalist blocks of the estate were all based on a shared, communal way of living and people instantly wanted their own home. Every home built these days is a detached home. You may not be able to fit a credit card between them and their neighbour, but everyone wants a detached individual home.[13]

As Chapman-Fox argues, throughout the 1980s the idealism that drove the New Town programme, and the associated provision of subsidized housing, was undermined by the drip-drip of discourses which depicted council housing, public transport and communal living as second-rate, socially and technically deficient. A popular urban myth circulated that Margaret Thatcher once described any man who used public transport aged over thirty a 'failure in life': whether she said this or not, the promotion of the right to buy, coupled with the stigma heaped onto social housing estates, portrayed the New Town ideals of the post-war era as absurdly anachronistic.

Nowhere seemed to exemplify this better than Runcorn's Southgate housing estate (Figure 4.4). Once heralded a unique experiment in deck housing, the 1,355-unit housing development, linked to Shopping City, was designed by award-winning architect James Stirling to evoke the Georgian squares and colonnades of Bath and Edinburgh but with porthole-like windows to symbolize the maritime heritage of nearby Liverpool. Completed in 1976, its unusual fenestration led to the estate being dubbed a 'concrete launderette'. From the offset it was associated with complaints about poor sound insulation and expensive heating: the decks and squares designed to promote community through strategies of spatial indeterminacy were shunned by occupiers used to private gardens.[14] Deck access and multi-level skyways made surveillance difficult and heightened anxieties about crime. As Hugh Pearman writes, the estate rapidly took on the mantle of a 'dumping ground', with families leaving and homes falling empty.[15] In 1989, when the Warrington and Runcorn Development Corporation was duly dissolved, the decision was taken to simply demolish the estate rather than commit the local councils to future maintenance and repair. Like Pruitt Igoe in the US, demolished after only twenty years, Southgate became emblematic of the failures of Modernism, albeit Pearman suggests the

Figure 4.4 *Interim Report, March 1979* (courtesy: Castles in Space) and the Southgate Housing Estate, Runcorn (Wikipedia commons).

destruction of the estate was ideologically motivated and far from necessary: refurbishment and regeneration might have allowed the homes to be improved at much lesser cost than the expensive demolition that occurred. Irrespective, they were ultimately replaced with 308 semi-detached brick-built bungalows in a low-density, privatized and individualized suburban design that represented the negation of Southgate's collectivist ambition.

The Southgate estate features prominently on the front of the first WRNTDP LP, where the music hints at the gap between utopian ambition and sometimes-dystopian outcomes. Tracks like 'Towns of Tomorrow' suggest a dissonance between the initial conception of communitarian living and the forms of alienation experienced by its inhabitants. However, Salvatore Dellaria has retrospectively argued tenant dissatisfaction with the estate was overstated: the state ultimately targeted it for demolition not because it had 'failed' but because it promoted forms of mutualistic socialism anathema to a Tory government invested in the cartographies of suburban privatism.[16] 'Right to Buy' was one of their key policies, and a major vote-winner, promising to make Britain a nation of homeowners. Chapman-Fox hence notes the convergence of postmodern architectural aesthetics and Conservative ideology, highlighting the gradual replacement of Modern housing by a postmodern vernacular:

> With these half-timbered houses in Warrington, it feels like everything has gone more and more regressive . . . Trying to do those ersatz Georgian mews properties all over the country is entirely regressive and part of me is thinking, how much does the architecture of the country actually feed into Brexit? How much of that 'It's got to be old to be good' mindset actually populates the narratives that helped fuel Brexit?[17]

WRNTDP's sonic exploration of the New Towns programme in the North-West contrasts the ideals of the past and 'the futility of the present' through an almost perfect alignment of faux 1970s aesthetics with a statement about capitalist realism. As he tells me, 'As much as the post-war developments are now widely derided, they genuinely set out to make lives for ordinary people better on a colossal scale', continuing by arguing 'No modern government has a thousandth of the ambition, drive or indeed obligation to the general public as those that established the post-war New Town programme'. Here, Chapman-Fox echoes Owen Hatherley's plea for reassessment of the much-derided architectures of Modernity: 'It was not utterly shit, and a significant minority of it was brilliant, and far better than anything built for most people since . . . we need a similar project now, and one which doesn't make the same mistakes'.[18]

In this sense, both the WRNTDP album *Your Community Hub*, released in the election year of 2024, and the accompanying single 'A Shared Sense of Purpose', seemed politically charged statements. On the cover of the latter was the community centre, designed by Peter Garvin, that stood at the heart of the Castlefields neighbourhood in Runcorn, a 2,400-home neighbourhood completed in 1972. This boasted a doctor's surgery, a Post Office, a laundrette, a pub, a supermarket, a primary school and a faith centre. As Chapman-Fox stresses, 'the idea was very much for these places to be like a village in their own right, with nobody's house more than five minutes' walk away from these facilities. Why would you not want that?' Here, WRNTDP's work draws a direct connection between the abandonment of neighbourhood principles in urban planning and the death of community. As he argues, any 'collaborative, community atmosphere seems to have gone now. It's just little things like knocking on your neighbours' door to make sure they're alright. So many people don't even know their neighbours' names. It's affected us all'.[19]

Tracks like 'A Brighter and More Prosperous Future' from 2023's *The Nation's Most Central Location* hint at the disappointment that was to follow the original optimism that accompanied the New Town programme in general and the plans for the economic revival of the North-West in particular. This is an elegy for a more hopeful era:

> To the planners and architects, it was very much a cause. No matter how misguided some of their designs may have been, they tried to plan and design the ultimate in affordable, modern communities, using the latest knowledge in many disciplines including architecture, economics and sociology. And although they made mistakes, their intentions were grand. The one thing they didn't plan for was neoliberalism killing off the entire ethos . . . The Thatcher revolution killed off the New Towns movement and many of the jobs and industries around Warrington and Runcorn. Anything that's happened since Cameron has been an acceleration of that.[20]

As Chapman-Fox notes, while the marketing of Cheshire as 'Britain's Most Central Location' seems almost risible now, at the time it was part of a determined national strategy to disperse office jobs from the overheated Southeast to more peripheral regions, albeit this was abandoned by the Tories when they began to promote property speculation in the capital by bankrolling London's Docklands in the 1980s. Chapman-Fox draws parallels with the political abandonment of the North in the 1980s and the failures of more recent regional development initiatives, arguing, 'We've had wave after wave of Northern Powerhouses and Levelling Up campaigns . . . they're all just hot air and empty promises'.[21]

Crumbling utopias

By coincidence, many of the ideas explored by Gordon Chapman-Fox are also interrogated on another album about a New Town in the North-West. The Magnetic North was a collaboration between Erland Cooper, multi-instrumentalist Hannah Peel and ex-Verve guitarist Simon Tong. Their 2013 debut LP was a symphonic tribute to Orkney, where Cooper grew up. Tong, in contrast, moved with his parents to Skelmersdale in 1984, two decades after it had been designated a second-wave New Town: like Warrington, it was intended as an overspill town for Liverpool. By the time Tong arrived, economic decline in Liverpool meant the projected population of 80,000 was drastically reduced, with crime, drug abuse and poverty widespread. The LP *Prospect of Skelmersdale* (2016) is named after a 1972 public information film about the work of the Development Corporation that vaunts an economic renaissance for the area, highlighting the companies – Thorn EMI, Courtaulds, Dunlop – attracted by a 'pollution-free' town providing a pleasant environment for workers.[22] Yet the LP's cover image by photographer Stephen McCoy, of a girl proudly cradling a racing pigeon at the rear of a 1960s housing estate, hints at more workaday realities, playing into myths of life in Northern towns.

Prospect of Skelmersdale is a kitchen sink drama set in the 1980s, punctuated by short samples from public information films from earlier eras. Musically, it veers from Belle and Sebastian-esque vignettes of domestic life in a New Town to baroque chamber pop and more epic synth-based, atmospherics. 'Cergy-Pontoise', for example, is a wistful, New Order-inspired instrumental that speaks of Skelmersdale's twin-town in the Île de France, ending with a single repeated refrain, Peel's voice defiant: 'I won't get down'. In 'Little Jerusalem', Peel refers to 'throwing my pills away', again alluding to the New Town Blues; in 'Northway/Southway', the New Town's gyratory road becomes a metaphor for the sense of being stuck in a failed utopia, Tong singing, 'I just want to step out of this world for a moment', the track fading to the voice of Richard Phelps, head of the

Skelmersdale Development Corporation, stating in 1972 that 'If people are not happy then we've failed'.

Prospect of Skemersdale does not repeat the familiar sonic tropes of hauntology but explores ideas of hope and hopelessness through an ambitious, almost cinematic account of life in a New Town. Tong confesses:

> Hauntology wasn't a term I was particularly familiar with when we first started. We were just instinctively trying to play with and draw inspiration from our own cultural memory and landscape and didn't know there was a name for it. We certainly found a rich seam of material with the idea of channelling ghosts or spirits who had lived in the places we were writing about. In Hilary Mantel's autobiography she talks about the spirits and ghosts she felt gave her ideas and with who she shared her everyday life. If you just give up to the idea there is some kind of matter and energy from the past (or the future) all around it can be very fertile ground for creating art.

Hannah Peel recalls 'I said to Simon that we should all go to "Skem" and see if anything came to us. He humoured us and wrote Erland and I a road map with a paragraph on each place . . . It was hilarious – truthful and a little bit ruthless'. As Tong tells me:

> We spent time there as a group – writing and recording – with the final goal of performing the album in the place we'd written about. We [*had*] a list of a dozen or so song titles that were actual physical place names which we would then write ideas around – like mini soundtracks – so it felt like each piece of music was directly rooted in the landscape. For the Skelmersdale album a lot of the track titles were the names of housing estates that had themselves been named after the farm/mining land that was there before the new town was built.

The resulting album was ultimately performed in Skelmersdale in a youth club, with McCoy's 1984 images projected on the wall alongside ones of the band pictured in the same locations (Figure 4.5). Peel recalls 'Skelmersdale has always had a lot of bad rep in the press about what happened there, but local residents are extremely passionate about their town, which meant that night of music was met with such energy and love – I've never felt that much excitement and fire in a room before. It was exhilarating'.

When performed live in Liverpool, the album was accompanied by a reading from author Frank Cottrell-Boyce describing the displacement of Liverpool families to Skelmersdale as akin to the movement of refugees. In an accompanying piece in *The Observer*, Cottrell-Boyce outlined the deficiencies

Figure 4.5 The Magnetic North. Courtesy: McCoy/Wynne.

of a town known for its endless roundabouts, fear-inducing underpasses and needle-strewn greenways, but suggested The Magnetic North album is altogether sunnier and more optimistic than this might suggest, capturing the hopes of the new arrivals to the town. He drew out the political dimensions of this by arguing that for all its failings, the New Towns embodied a 'discipline of dreaming' all but vanished from contemporary politics. In his words, 'We lived through a utopian project and we saw utopia crumble. But that doesn't mean the idea of utopia is not useful'.[23]

Here, both The Magnetic North and Warrington-Runcorn New Town Development Plan have turned a seemingly esoteric interest in the New Towns of the North-West into music with wider resonance, especially for those who grew up in the New Towns or the expanded towns of the 1970s. As Chapman-Fox explains to me, 'I've been contacted by so many people who grew up in New Towns to tell me about their experiences, and how it sounds like their youth. It's really struck a nerve'. But even for those unable to directly relate to the experience of living in a New Town, music that speaks to the futility of the present by looking back to the utopianism of the past is intensely appealing. Gordon Chapman-Fox suggests that until he conceived WRNTDP, he seldom sold more than thirty copies of any release: in contrast, *The Nation's Most Central Location* broke into the top thirty UK physical album sales chart, and gigs are often sold out. His architectural hauntology has certainly struck a chord.

In his obituary of popular Modernism, Mark Fisher suggests bulldozed Brutalist buildings are a sign that the promised future did not arrive.[24] But in a somewhat surprising turn, those remaining Modern buildings often dismissed as ugly, proletarian or workaday in the late twentieth century are being revalued in the twenty-first. Here, the music of WRNTDP, The Magnetic North and Town and Country stands alongside a slew of photobooks celebrating Brutalist developments as diverse as London's Centre Point, Birmingham New Street Signal Box and Newcastle's Byker Wall. The best-selling photographic gazetteers of Brutalist architecture produced by Simon Phipps are acclaimed as concrete poetry, while the homeware company People Will Always Need Plates sells tea towels, mugs, coasters and tote bags with images of London tower blocks emblazoned across them. While this could all be dismissed as a middle-class recuperation of Brutalism that severs it from its social function, those producing these artefacts are at pains to connect Modern architecture to the wider public cultures of the 1960s and the social agenda it promoted. The music of WRNTDP, The Magnetic North and Town and Country – as well as Black Zephyr's (2023) tribute to the M6's Forton Services, 'Pennine Tower', Carlisle City Council's (2024) *The Lanes Re-development*, or The Twelve Hour Foundation's (2023) *Hexagonal World* – all remind us that Modern planning and architecture was conceived to privilege people over capital, based on an ethic of inhabitation that actively encouraged social and cultural co-operation of a type that has long since disappeared.[25] At a time of acute housing crisis in England, especially in cities pockmarked by ill-conceived regeneration schemes lining the pockets of private developers, it's easy to see why so many are haunted by the landscapes of urban Modernity.

Further Reading

Craggs, Ruth, Geoghegan, Hillary, & Neate, Hannah (2013). 'Architectural enthusiasm: Visiting buildings with the Twentieth Century Society'. *Environment and Planning D: Society and Space*, 31(5), 879–896.

Gold, John (1997). *The Practice of Modernism: Modern architects and urban transformation, 1954–1972*. Routledge.

Grindrod, John (2018). *How to Love Brutalism*. Batsford Books.

Hall, Peter (1998). *Cities of Tomorrow: An intellectual history of urban planning and design in the twentieth century*. Blackwells.
Hatherley, Owen (2009). *Militant Modernism*. John Hunt Publishing.
Mould, Oli (2017). 'Brutalism redux: Relational monumentality and the urban politics of brutalist architecture'. *Antipode*, 49(3), 701–720.
Pearman, Hugh (2010) 'What went wrong at Runcorn'. https://www.architectmagazine.com/design/urbanism-planning/what-went-wrong-at-runcorn_o
Thoburn, Nick (2018). 'Concrete and council housing: The class architecture of Brutalism "as found" at Robin Hood Gardens'. *City*, 22(5–6), 612–632.

5

The Sound of the Edgelands

While the English rural is reflected in the musical traditions of the pastoral, evoking serenity and beauty, and the city through the slurry hauntronica of Warrington-Runcorn New Town Development Plan or Pye Corner Audio, most of us live and die not in town or country, but the land in between. Indeed, while an estimated 20 million live in the country's major conurbations (35 per cent), compared with just four million (8 per cent) in villages, most in England live in smaller towns and cities, or at the urban-rural fringe. The latter denotes the interpenetrating urban and rural land uses associated with urban sprawl: it's a landscape that's not quite rural but not completely urban either: a truly 'sub-urban' landscape where green spaces are mainly in the form of sports pitches or allotments rather than field and forest.

While suburban living promises the best of town and rural living, and easy access to green space, Marion Shoard famously described this meeting point of city and countryside as the *edgelands*:

> Between urban and rural stands a kind of landscape quite different from either, a dismal landscape of rubbish tips and warehouses, superstores and derelict industrial plants, office parks and gypsy encampments, golf courses, allotments and fragmented, frequently scruffy, farmland . . . These heterogenous elements are arranged in an unruly and often apparently chaotic fashion against a backdrop of an unkempt wasteland frequently swathed in riotous growths of colourful plants, both native and exotic . . . In our own age, this zone has expanded vastly in area, complexity and singularity. Huge numbers of people now spend much of their time living, working or moving within or through it. Yet for most of us, most of the time, this mysterious no man's land passes unnoticed: in our imaginations, as opposed to our actual lives, it barely exists.[1]

In relation to music and landscape, the latter observation is remarkably astute: the edgelands are a non-place realm that arguably lacks a musical identity, seldom having inspired or provoked musical reaction.

But following the trend to scrutinize nature in more mundane places, influenced by Richard Mabey's (1973) *The Unofficial Countryside* and Farley and Roberts' (2012) *Edgelands*, there has been a recent upturn in musicians drawing inspiration from the interface of town and country, embracing the unruly, hybrid and paradoxical nature of the urban fringe. The COVID-19 pandemic added impetus to this, with the lockdown in England limiting people to limited daily exercise in the local area, encouraging some to wax lyrical about the restorative qualities of spaces rarely spoken of in those terms: parks, canal towpaths, and even abandoned industrial sites became sites where lockdown-weary residents found respite from the daily deluge of data on coronavirus cases and deaths. Here, many people's experiences during lockdown corroborated Mabey's claim that edgelands are a 'therapeutic landscape', offering healing, excitement, beauty, transformation and a change in the pace of life.[2]

Lockdown was a time of contradiction. While the boundaries of daily experience effectively closed in, many developed a heightened and appreciative awareness of their own locality. So while lockdown starved musicians of opportunities to perform in public, and threatened the viability of many venues, for many musicians it inspired new projects born out of an enforced engagement with their surroundings, no matter how mundane. An example is provided by Matthew J. Saunders, formerly of Magnétophone, a band emerging from the same Birmingham post-rock scene that spawned Broadcast. Now recording as Twilight Sequence, Matt decamped from Birmingham to Somerset for a change of scene and during lockdown began to build a modular synth. For Matthew, the brief periods of daily exercise permitted under lockdown provided the prime inspiration for his subsequent music-making:

> During lockdown I spent time in the woods and green lanes near the studio: it was like walking through the nature books and programmes of my childhood. When I stumbled upon old and antiquarian natural world books in the local charity shops, I was inspired by them as well, especially an encyclopaedia by John Hammerton which was called *Outline of Nature in the British Isles*, as well as Ladybird nature books and other specifically British natural world books.

The resulting album, *Outline of Nature* (2023), is a 'voltage-controlled outpouring of love' for the local environment, coloured by Matthew's view that 'as a species we have destroyed much of our habitat and the wildlife that once thrived in it'. On this album, the tracks are not just inspired by the landscape but directly *feed* on it. Matthew tells me that each track was a response to his daily lockdown walk:

> I felt connected to nature on the day as I wrote it, part of the rotating season at that very point in time when I tuned the oscillators and the rain started to fall on the studio . . . because I recorded everything without overlays, each track was a live performance fed by the environment, and captured few moments of a particular, single and real moment in time.

He continues:

> When I switch the studio on and fire up the modular, I feel like I'm interacting with the same energy, the same life-forces that the natural world is created, and grows, from. My studio is on a hill, and though we're in a residential area, it's pretty green and there are trees all around. There's a little patch next to the studio that has grown wild, there's also a substation next to it, and it really feels electric there, I feel like I'm tapping into the electricity of the place and I'm guiding it with voltage controlled oscillators, envelopes, amplifiers and filters, coaxing it into the audio spectrum and ultimately into the emotional spectrum because that's where we operate most thirstily, and that's how nature resonates in us, on that emotional and wet, yet still electrical, level.

The track 'Rotating Seasons' is accordingly built around a repeated tuneful electronic keyboard riff, and 'Birds in General: and the Rook' (named after another of John Hammerton's books) throbs with energy as it builds to a crescendo. Twenty-minute stand-alone single 'Trees in General and the Larch' likewise builds slowly, melodic patterns layered intricately as if to reflect the structures of a tree (Figure 5.1). Reflecting on this, Matthew notes: 'The woodlands that helped me write the LP may be a retail park in twenty years' time, and that's a likely future I don't want . . . If my music could shape anyone's disposition to the world, I would hope it could be something really simple, like someone might stand for a moment and look – I mean *really* look – at an oak tree, and feel the wonder and the power of it, vibrate with it'.

One half of electronic act ISAN, Robin Saville, is another artist whose work has always been receptive to the sounds of nature (Figure 5.2). His 2013 album *Public Flowers*, for example, combines quiet pastoral field recordings and simple, cascading synth and metallophone melodies alongside often-skittering electronic percussion. Influenced by the minimalism of Piero Milessi, Steve Reich and Terry Riley, Saville's work builds complex and fragile compositions from simple and repetitive melodies, mirroring how the beauty of nature is constructed

Figure 5.1 The tree inspiring the Twilight Sequence's 'Trees In General and the Larch' artwork. Courtesy: Matt Saunders/Zeke Clough.

through regularities of form and structure. Tracks include 'The Bryophyte Annual Picnic', built around a loping bassline and scampering hypnotic bleeps, and the 'Long Walk from Sallowes Church', a ten-minute track which starts with the noise of bird and insect life, before gentle xylophone melodies and charming

Figure 5.2 Robin Saville. Courtesy: the artist.

synth phrases meander around each other to beguiling effect. Tellingly, the title *Public Flowers* comes from E. Nesbit's (1906) book *The Story of the Amulet* in which a group of children are magically transported into the future, presented as a wonderfully clean and tidy place where the citizens learn to recite:

> I must not litter the beautiful street,
> With bits of paper and things to eat;
> I must not pick the public flowers,
> they are not mine but they are OURS.

A keen allotment holder who also works in a farm shop, Robin Saville notably paid for remixes of the tracks on *Public Flowers* (released on the *Hybrids* EP) with homemade jam, a gesture that speaks of his desire to promote forms of social and environmental sustainability: he also tells me that too many today live in a 'solipsistic bubble' that blinds them to the possibilities of multi-species life.

Saville's 'lockdown' album *Lore* (2023) further evidences his interest in everyday nature, its gentle rhythms and glitchy electronics built around field recordings made in the vicinity of his Suffolk home:

> The field recordings in *Lore* are very much a record of a five-mile radius of where I live. And that was that was forced by lockdown. But that doesn't mean

> that if it weren't for COVID-19, I'd have been out making recordings in Mexico or something like that. I've always kind of mostly do my recording in the UK anyway, cause it's you know, it's where I live and it's where I'm trying to be out in the environment as much as possible. The whole notion of lockdown art is a bit hackneyed, even a bit boring, but I was forced to explore my local environment and you know, take what I could from that environment.

Saville stresses the field recordings were no mere embellishment, with the music built around the field recordings rather 'than one just being a kind of decorative aspect on top of the other'.

Lore is then a notable example of 'coronamusic' inspired by an enhanced awareness of the local environment, simultaneously indicative of the therapeutic dimensions of nature at a time of social anxiety and collective stress. Notably, during the spring of 2020 Saville continued to work in his local farm shop (which remained open during lockdown), finding the experience of working while most people were furloughed intensely stressful. As he recalls, 'when I wasn't at work, I really needed to be kind of out and de-stressing'. Field recording was a way of releasing this tension:

> So I was out making recordings. There's a recording on there of a swarm of bees. I was just out on a bike ride, and I happened to notice this swarm. So I cycled home, grabbed the recording equipment and came right back. Sellotaped it to the end of a long stick, stuck it as close to the swarm as I dared and hoped for the best!

Saville's emphasis on the forms of nature encountered in the urban fringe is something he suggests does not feel is any more 'valid or important' than any other nature, but at the same time he argues 'it's to be celebrated and it's wonderful because it's magic when it's encountered in a very mundane setting'. The titles of the 'musical miniatures' on *Lore* reflect this mundanity. Opening track 'Judith Avenue' for example, is named for a road near Saville's home where he recorded the song of nightingales during lockdown:

> It's a road that starts off at one end, all quite neat and tidy and it just sort of fades off into scrub. It just sort of ends as wasteground. And if you jump on your bike or if you have a warm evening in May, jump on your bike ride from one end of the road to the other, leave behind the streetlights and here there are nightingales that have just flown here from Africa and . . . How could I not want to share?

For Saville, the incorporation of motifs representing the natures encountered in the mundane surroundings of his home in a Suffolk town is an insistence we should celebrate the wildlife that surrounds us, but not downplay the way anthropogenic activity threatens it.

More-than-human Manchester

During lockdown, the absence of traffic on the road, the lack of planes overhead, and the closure of nightlife allowed people to hear things that were always there, yet usually silenced by the dominant, louder sounds of everyday life.[3] Auditory experiences outside the 'normal' range of experience became available like never before, with COVID-19 lockdown conceptualized in some quarters as an *anthropause* that meant it was no longer necessary to use expensive probes, sensors or microphones to hear nature. This 'recalibration of attention' not only led to a generally heightened sonic awareness of the animals that live alongside us – and related practices of bird-feeding and butterfly-spotting – but also inspired many musical projects. *Island of Noise*, for example, by Jack Cooper's Modern Nature, came in two versions, one instrumental (*Island of Silence*) and one replete with lyrics contemplating the destruction of the natural environment, the title reminding us that in England it's difficult to ever escape the noise of traffic.

Lockdown also inspired experimental work using remote recording devices and sensors, albeit much of this rejected 'bioacoustic' approaches regarding the sound of nature as scientific 'data'. Instead, projects were often designed to encourage a sensate awareness of the role of the more-than-human in the making of our world, as A M Kanngieser argues:

> By listening to places and spaces, it is possible to discern uneven, often-obscured, and gradual processes, such as extinction and dispossession. Sound is not just about hearing and responding, or communicating. It is about becoming aware of registers that are unfamiliar, inaccessible, and maybe even monstrous; registers that are wholly indifferent to the play of human drama. Sound is not only of the human, it undermines human exceptionalism: everything vibrates on some frequency and is touched by vibration, regardless of how imperceptible to human sensibility this might be.[4]

A rich variety of 'ecological sound art' hence emerged during lockdown, often using the new-found clarity of the suburban soundscape to encourage a more compassionate co-species existence. This was not necessarily about presenting certain natural sounds as superior to those usually associated with urban life (as in Murray Shafer's aestheticization of the natural soundscape) but *de-centring* the human to recognize that our surroundings are made by all number of human, non-human and more-than-human actors.[5]

Hayley Suviste's *Edgeland* is exemplary here, weaving together recorded conversations, composed music and binaural field recordings from a ten-acre reclaimed greenspace on the boundary of Trafford and Manchester, Ryebank Fields (Figure 5.3). Far from being pristine nature, Ryebank Fields stands on the site of a brickworks closed in the 1930s that became an unofficial landfill

Figure 5.3 Hayley Suviste, Ryebank Fields. Courtesy: Rob Herissone-Kelly/Hayley Suviste.

site for builder's rubble. Reclaimed in the 1970s, it was subsequently gifted by the City Council to serve as sports pitches for Manchester Metropolitan University. Ryebanks Fields were then abandoned by the University in the 1990s in favour of a larger sports complex elsewhere, though the University remains the landowner.[6]

Since their abandonment, Ryebank Fields have been slowly taken over by nature, becoming rough pasture with thickets of scrub alongside some wooded areas, including clumps of willow, rowan and sycamore. Other plants present include dogstail, reed canary grass, perennial ryegrass, cocksfoot, red fescue, brome, rosebay willowherb, elderflower, meadow buttercup and rare bee orchids, as well as blackberry bushes that attract jam-makers and are turned into wine by a local producer who sells them from his barge. Like many greenspaces at the urban margins, Ryebank Fields is also a popular site for dog walking, family picnics and morning joggers. During COVID-19 lockdown, it took on the role of an outdoor community centre, its ordinary wildness making significant contributions to health and wellbeing.[7] Suviste tells me she found herself naturally drawn to Ryebank Fields, and that while 'confines of the city felt increasingly claustrophobic, this area provided a refuge where I could immerse myself in nature in a way that more manicured city parks did not, something crucial for grounding myself and alleviating the mental health challenges posed by the lockdown'. Another voice we hear on *Edgeland* echoes this:

> That was the really good thing about lockdown: it gave us time and the Fields were full of people. Paths came back that hadn't been walked for years, but they were there. There's an old boundary along the park, and that was being used more than it's ever been. The value of nature soared during lockdown, it is all we had. Everything was amplified: the leaves and colours!

Ryebank Fields is presented here as both a secret and accessible environment, a wasteland *reinvigorating* the community. Local volunteers have helped maintain this de facto therapeutic space via regular litter picking, planting cornflowers and poppies to encourage pollination and even burying acorns to encourage tree growth (one dogwalker planted 300 at the turn of the millennium, resulting in over 100 oaks on site). But unlike those spaces protected by Green Belt legislation, Ryebank Fields doesn't enjoy any formal designation as a green space in local planning documents and has no particular archaeological significance either. As such, the site was slated for development by the University in 2017, with Step Places and Southway Housing Trust the preferred developer for a proposed 120-home community. However, when beginning clearance work on the site, asbestos was found among the rubble, including some thought to have been dumped during the construction of Manchester's Arndale Centre in the 1970s. Fenced off on the pretext of testing for more asbestos, the park was nonetheless occupied by Extinction Rebellion protesters in April 2021 who constructed a community camp based around an 'eco-house' built of wooden pallets, challenging the University's claim the development would result in biodiversity gains. One protester – Steven – argued: 'By protecting this field, we're giving species in the UK a fighting chance of survival. We need to have habitats like this, where, unlike man-made parks, plants and animals are just left alone to do what they want and thrive'.[8]

Hayley Suviste's *Edgeland*, sold to support the 'Save Ryebank Fields' Fundraiser, involved a mixture of field recording and in situ composition:

> I've begun recording myself playing instruments like the clarinet or shruti box outdoors, improvising and responding to the environment in real time. I experiment with different drumsticks to create rhythmic and textural sounds, capturing these interactions from various perspectives by placing microphones in diverse locations (right next to the instrument, at a distance, or even hanging from a tree). Additionally, I use different types of microphones, such as contact microphones embedded in tree trunks or the ground, geophones, and electromagnetic microphones, to gather a wide range of sounds from the same space. When I bring these recordings back to the studio, I layer and edit them, creating a journey for the listener by integrating multiple perspectives from the same time and place.

Over ambient drones and improvised synth melodies, this musical journey includes extracts from interviews with locals. Towards the end of the end of the twenty-five-minute piece, one speaks of the value of the space for their children:

> Because we live in a city, children need to have the opportunity to just play in these wild spaces, it's invaluable. We just don't have many of these spaces left. With many a young child growing up in a city that has no real engagement with nature on a day-to-day basis, how is that child then expected to care for its environment? How is that child then going to learn about nature?

These thoughts echo the language of the great anarchist writer Colin Ward, whose *Child in the City* (1978) offered a critique of social attitudes towards urban play and children's safety.[9] As he put it, children need urban 'wild places' to develop an environmental education, something that cannot be obtained in sterile urban spaces where opportunities for play are limited. To emphasize this, during the final moments of *Edgeland* the adult voices and soundbites fade out, replaced by the sound of children playing and birds quietly chirping, stressing the value of the space in both the social and environmental futures of the city. The work is hauntological in the sense that this land is threatened, the liveliness of Ryebank Field in the recent past contrasted with its current dormant status, a space held in eerie suspension while awaiting redevelopment.

More generally, Manchester has been one of England's cultural and economic powerhouses in the twenty-first century, its post-industrial revival underpinned by the attraction of the 'creative class', the relocation of the nation's media industries and an enviable reputation for music and nightlife. It is also a student city, with five higher education institutes boasting nearly 100,000 term-time students. All these factors have driven demand for property, with the city having witnessed explosive urban growth, and the local authorities of Manchester, Salford and Trafford frequently providing direct and indirect subsidies to developers, particularly those pushing a sustainable agenda. One such developer is Peel Waters, who are currently building 2,700 homes on the 25-acre Pomona Island site, wedged between the River Irwell and the Bridgewater Canal. Peel describe this site as underdeveloped brownfield land, arguing the conversion of 60 per cent of the Island into public park, with planting schemes along the canal towpath, will transform this 'wasteland' into a green lung for the city. But like the users of Ryebank Fields, some activists are resisting what might be termed its 'ecogentrification', suggesting the area is already bountiful in terms of flora and fauna, and that the redevelopment will destroy any hope of developing a genuinely green urban oasis:

> Although to a pedestrian the island is almost completely isolated, with only one dedicated footbridge, it takes just minutes to reach Pomona on foot from Manchester's city centre. And yet it seems worlds away from anywhere. There's a brackish, almost sand-dune quality to the hulks of upturned land, formed by the uprooting of mature trees by current landowners Peel Holdings. Over these old roots, new saplings have laid down their own. Moss and lichen carpet the corners of cracked paving along the periphery of the site, and all around Pomona is a defiant hive of life that has thrived on the urban decay that came before.[10]

It is easy to see how developers are able to discourse the area as wasteland: former dock structures are strewn with graffiti, piles of fly-tipped rubbish and soil contaminated with industrial by-products of days gone by. But precisely due to its neglect, this left-over urban space has more species richness than the traditional parks and public spaces of the city.

In this setting, Hayley Suviste has evolved an augmented soundwalk which re-imagines Pomona Island's history through field recordings, electroacoustic music, oral histories and archival material that's accessed by the walker as they proceed through the area. We hear a canalboat gently chugging past, giving way to the amplified sound of iron railings being struck – connoting the former docks built on the site in the nineteenth century – and then the noise of a larger vessel, a boat possibly exporting Manchester-manufactured goods to the Americas in the early years of the twentieth century. The track then gives way to the contemporary sound of geese and ducks, as a train trundles towards Piccadilly station, just a few hundred metres distant from the Island.[11] Presenting Pomona as a rich 'urban oasis', layered with histories and memories, Suviste's piece uses resonant sounds to evoke the forms of non-human life that have taken root in this landscape, always emphasizing the proximity of the gentrifying city. As Suviste argues, 'A lot of people with field recordings want to block out sound and make these perfect nature birdsongs, but I think to include human sounds in cities is important, you can hear politics in it'.

The ecology of the urban wasteland is also an important theme in much of the 'new nature writing', where interest in ruderal ecology is now prominent. Referring to the heterogenous forms of life emerging in urban ruinscapes, anthropologist Bettina Stoetzer argues that the perspective of ruderal ecologies allows purchase on how nature often resists human subjugation, and exhibits a real sense of agency. As Stoetzer describes, 'weeds' force themselves through the cracks in concrete, with spontaneous forms of life taking root in spaces that at first glance appear inhospitable. Typically, the species best able to thrive in such urban contexts are invasive rather than 'native' species. From this perspective, Stoetzer suggests ruderal ecologies show that biological life, community identities, and strategies of survival are never authentic but are always situated within histories of migration and disturbance:

> Following stories about rubble, gardens, and other human–environment relations in the city means looking at how the ruins of contemporary capitalism and nation-making become our very habitats . . . Rubble plants grow from seeds carried on the boots of migrants and refugees, and they may remain dormant potentials in a city's fabric.[12]

Stoetzer's explorations in Berlin resonate with the rubble stories of Manchester, where histories of industrialization and connections to Empire have produce a de-industrial landscape whose inherent complexity and beauty is ignored in the city's promotional brochures and planning documents which instead extoll the virtues of manicured green corridors and rooftop gardens. Suviste's musical interventions hence point to the dangers of ongoing ecogentrification, reminding us that ruinscapes that have a certain amount of healthy disorderliness or mess are not only good for wildlife habitat and the health of the planet, but for our souls as well.

Steel city sounds

Across the Pennines, Sheffield-based composer and musician Liz Hanks explores similar themes in her beautiful and elegiac 2023 album *Land* (Figure 5.4). Having worked as a session musician for over twenty years, performing

Figure 5.4 Liz Hanks. Photo: author.

with artists including Paul Heaton, Liam Gallagher and Richard Hawley, during lockdown Hanks was inspired to explore her local environment and work on the project that resulted in her debut LP. As she describes, 'boredom is one of the most creative forces – there are benefits to doing nothing . . . you start thinking in novel and productive ways'.[13] Performed live with cello, loop pedal and iPad, *Land* combines field recording with ambient classical and folk motifs to soundtrack the Meersbrook area of south Sheffield. This is a now-suburban area once known as Rush Dale, the valley through which the Meer's brook flowed into the River Sheaf (the principal river that gives the city its name). This stream was culverted in the 1960s and is now hidden from view to the rear of Albert Road off the A61, the stream only readily accessible where it rises in the woodlands behind the Gleadless Estate. Developed by Lewis Womersley, architect of the more celebrated Park Hill estate, this 1960s Estate was developed as a mix of maisonettes, houses and high-rise flats accentuating the contours of 'a beauty spot considered too steep and north-facing for development in the 1930s but purchased in desperation in 1952–53'.[14] This means that this stretch of the valley is now almost entirely urbanized, but vestiges of the rural past remain in the form of place names, traces of former farm buildings and the green wedges that remain despite decades of development at the edge of Sheffield.

Hanks' album is rich and expressive, using the cello to evoke different moods and textures, as she explains:

> My music is instrumental, free from words or lyrics. I find the absence of language to be an advantage, enabling me to summon the essence of a place without the confines of detail or visual representation. I believe incorporating field recordings enhances this sense of place, while the cello – sometimes improvised, at other times composed – deepens the connection to the past. I want this music to revive the history of this urban landscape, inviting us to catch a brief glimpse of the past and feel the echoes of what once was.

Land is accompanied by an app-based walk, with commentary by local poet Sally Goldsmith guiding the listener from Meersbrook Park, upstream past the Heeley allotments into Carr Wood and then back down Lees Hall Avenue, a loop of some three miles. Along the route, the album offers instrumental accompaniment, encouraging the walker to stop and appreciate how nature and culture have long combined to create a distinctive sense of place. The track 'Carr', for example, speaks of the origins of the ancient oak and hazel trees that were once coppiced for wood to power the furnaces of Sheffield ('kjarr' being a Norse place name signifying a thicket or copse). The track 'Keld' (or spring) samples the liquid trickle of the stream itself, which now constitutes the divide between Meersbrook, Heeley and Gleadless districts, but was once the boundary line between Yorkshire and Derbyshire, and before that, between the Anglo-Saxon kingdoms of Mercia and Northumbria. Other tracks are subtly built around field recordings made by Alisdair McGregor. The sound of distant traffic, for example, is heard at the beginning of 'Ride', a track about the lost footpaths that once criss-crossed the Dale, while the crunch of footsteps on gravel pathways reminds us of the many who walked these paths before us. On the app, we hear the names of those who used to live in the local streets – grinders, cutlers, shopkeepers, and wood merchants – and are encouraged to explore the former orchards of Lees Hall, built in the 1500s but demolished in the 1950s. We hear the sampled sounds of sawing on the 'Kerf' (the width of a saw cut), and are told of Arnold Laver, the Meersbrook resident who became proprietor of the largest timber merchants in the city, having started his business working in the ancient and still-standing Carr Woods.[15]

Land is influenced by psychogeography as much as by hauntology, a tracing following the psychic contours and desire paths of a city laid down by the successive generations of residents who lived on, and with, the land. Hanks explains:

> I was captivated by the fragments of information I discovered about the lost valley, an area that was once rural idyll. Yet I was also fascinated by the gaps in information about this place, which was not significant enough to have made a mark in history. These gaps enabled me to create a more 'dreamlike' view of the past. Today, I see the local woodlands, parks, and green spaces as pieces of a jigsaw puzzle, revealing glimpses of the region's rural heritage. Although the places featured in my work might appear insignificant in today's context, I believe any piece of land possesses a unique and captivating natural history.

In this way, *Land* focuses on a remarkably ordinary urban landscape whose forms nonetheless show the traces of natural forms and processes too often overlooked. As Hanks argues, 'this is something people in modern society have become detached from: our cities feel so permanent, with solid roads and pavements, towering buildings and so on, but I am giving the message that the natural world was here first and should be respected'.

Empty houses, overgrown towpaths and old railway lines

Land is an exquisite album that straddles modern classical, electronic music and acoustic ecology, engaging with the close but largely uncharted spaces of the suburbs. It offers a sonic topography alive to the legacies of the more-than-human world, registering the way the country and the city bleed into each other. Inspired by Richard Mabey and others, there are many other artists, writers and performers currently documenting this edgelands.[16] Much of this work disrupts the neat binary between nature and culture, focusing on the intimate relationship between landscape, place and people.[17] It frequently tells a story that's personal *and* political, using autobiography as a tool for environmental critique, reflecting on both the promises of suburban living and their denial.[18]

Working in this rich vein, epic45 have created a distinctive soundtrack for the English edgelands over nearly three decades of musical exploration. Their music straddles town and country, melding delicate pastoral pop and discordant experimental noise to explore the human dramas and dilemmas of the urban edgelands. Since 1995, epic45 has been an evolving project based around the nucleus of Rob Glover and Ben Holton, two school friends from Wheaton Aston, Staffordshire, a village just beyond the green belt surrounding Wolverhampton and the West Midlands conurbation (Figure 5.5). Both bonded at school over a mutual love of the Pet Shop Boys, the paranormal and weird TV, their shared

Figure 5.5 epic45. Courtesy: epic45 / Anthony Harding.

musical interests leading to a series of singles and albums in the early 2000s on a variety of indie labels, including Make Mine Music, a Birmingham-based imprint whose roster included Yellow6, Avrocar and July Skies.[19] The band's early material is firmly in the post-rock idiom, with the influence of Mogwai very evident in tracks that exhibit a quiet/loud dynamic. Slowly, however, an interest in topography became more prominent in their evocation of atmosphere, their adolescent post-rock giving way to a more melancholic 'landscape music', with vocals added on a more regular basis.

It was when guesting on July Skies 2004's album *The English Cold* that epic45's overt interest in hauntology became evident, with the album finding the ghosts of WW2 at the airfield of Wheaton Aston, as Ben recalls:

> Both Rob and I shared a vivid interest in the paranormal as kids and often freaked each other out with weird stories and strange images from books like the Usbourne 'Mysteries of the Unknown' tome. But as I grew older I found myself less and less convinced there is anything other than our own minds that are strange and that we happen upon 'the uncanny' as we process certain situations or places. I still believe certain places have certain 'effects' on our psychology though, if you are that way inclined. One instance is when Rob, Antony Harding (of July Skies) and I visited the WW2 airfield near Wheaton Aston and explored what remained of the firing wall. We all 'felt' something as the light faded and the air chilled. We fell silent and split up as we looked around the ruined wall. When we returned, we set up the recording equipment and made what became 'Waiting To Land' on the WW2-themed July Skies album, *The English Cold*. It was a powerful experience.[20]

'Waiting to Land' is a sparse six-minute track where Ben Holton contributes guitar: crystalline notes hang in the air, spaciously echoing above washes of synthesized sound. Likewise, other tracks on the largely instrumental album, including 'The Countryside of 1939', 'Lost Airman', and 'Empty Skies', serve as a requiem for lost youth, the contemporary landscape bearing witness to the sacrifice of those lost to war. Remarkably, sixty initial copies of the album came complete with a fragment of a former Second World War airfield, chipped off by hand.

This nostalgic infatuation with landscape history subsequently informed the making of *Drakelow* (2006), an album where electronic sounds were to the fore, degraded tape loops and synth chords evoking the labyrinthine tunnel system under the Worcestershire countryside that was designated the regional seat of government in the event of a nuclear war. Like other musical acts before and since, this saw epic45 tapping into the nuclear uncanny by examining the ruins of the Cold War era that so haunted those who grew up in the 1970s and 1980s.[21] This 'bunkerology' was, as Ben notes, epic45's most obvious engagement with hauntology, something that appealed to them both aesthetically and theoretically. As he recalls, the band found the early releases on Ghost Box Records particularly inspiring, and at one stage Baron Mordant (of Mordant Music) was due to remix *Drakelow* in the style of his *Travelogues* series of hauntologically inspired sound collages.

But, emerging from below ground, epic45 subsequently eschewed the relict landscapes of war to explore everyday life on the urban fringes of the Midlands, where the fields sit beneath pylons and litter often chokes the hedgerows. As Ben tells me, it is this singular focus on the urban hinterland that has subsequently informed epic45's distinctive sound:

> We have never been part of any scene and never lived anywhere where that would even become a possibility. From the scruffy, non-picturesque village where we grew up in Staffordshire, we now live in the outskirts of the city, the suburbs. I write from that perspective and increasingly so. I've always thought that the Midlands is like one big 'edgeland' in a way. I see what we do as epic45 as 'edgeland music'.

epic45's work hence explores the entanglement of biography and place in the edgelands, something reflected in the band's music and lyrics, as well as the photobooks and images that have accompanied their many releases. The interplay of the music and imagery is central in the compositional process, with Ben often using photography as an inspiration for 'painting' with sound.

Tellingly, 2005's *Long Walks* EP contained a track 'For Virginia Astley' that offered a faithful imitation of the sounds of her 1982 pastoral album *From Gardens Where We Feel Secure*, complete with field samples, signalling the direction the band's music was moving in. But it was the *England Fallen Over* EP (2005) that encapsulated the edgelands aesthetic which epic45 has subsequently polished and refined. The EP cover featured the cooling towers of Ratcliffe-on-Soar power station, Nottinghamshire, a far from conventional representation of Englishness, revealing the band's increasing interest in the liminal landscapes of the Midlands. For one reviewer, the EP's combination of 'gentle guitars, sparse piano trickles, and measured drumming' created a sense of 'pastoral motion, like a train travelling through some half-remembered field'. However, the same reviewer noted the more ominous drones that intersperse these fragile moments, 'looming overhead the way the smoke-belching cooling towers loom on the cover'.[22] This ambivalent representation of the edgelands was undercut by lyrics depicting it as the realm of thwarted ambitions and fading memories. The lyrics of 'Swerving to Avoid Falling Leaves', for example, nostalgically dwell on the dashed hopes of youth:

> That day by the old school between terms
> What was left of summer keeping us warm
> We can sit and make plans
> That we know we'll never see
> Through to completion.

Here, Ben and Rob drew overt inspiration from growing up in Wheaton Aston, recalling views of 'South Staffordshire's farmed undulations, Wolverhampton's high rises, and the radio tower dominating Cannock Chase's woodland'. As they explain it, something 'immense' always appeared to be happening 'in the distance along the hazy horizon', encouraging them to consider the relationship between their immediate surroundings and events in wider society.

May the Heart be Your Map (2007) played on the importance of the countryside beyond the urban fringe, songs riffing on the turning of seasons and the cycles of nature, the title track inviting us to travel the paths less trodden (We'll take the back roads today / We'll navigate the lanes alone). On 'The Trees and Lanes', whispered words drift over a delicate pastoral guitar melody:

> Church spires spying through the trees,
> Sunlight through the dying leaves
> Are we the last survivors, just us, the trees and the lanes?
> Can we re-learn to think for ourselves
> When there's no place like home anymore?

On 'Summer's First Breath', a shuffling breakbeat and electronic rhythms are slowly layered with plaintive guitars and a hypnotic bassline that builds and builds, the accompanying video showing two schoolboys playing in fields, equating the slow erosion of the countryside with the loss of childhood:

> Church bells over fields
> And hope in our heavy hearts.

Don't just sit there, look outside
Leave the house far behind.
This world's a mess
But we'll stumble through
Windswept and weatherworn.

Word magazine later picked this as one of albums of decade, Jude Rogers suggesting 'their ghostly, fingerpicked guitars, sampled found sounds, memories of shoegazery and dusty electronica gave the Black Country its very own Boards of Canada'. An accompanying 12-inch of remixes from the album appeared as the first release on epic45's Wayside & Woodland label, later home to numerous epic45 side projects, including My Autumn Empire (Ben), Toy Library (Rob), The Balloonist (Ben), Birds in the Brickwork (Ben) and Field Harmonics (Rob with Bryony Williams). The ethos of Wayside & Woodland Records has been described by Holton as concerned with 'degradation of memory, decay, change, loss, the beauty of abandoned buildings and places, post-industrial wastelands, "edgelands", "unofficial countryside", grass verges, and woodland between housing estates': precisely the sort of concerns that coalesce where rural Staffordshire meets the urban West Midlands.[23]

2009's *All the Empty Houses* was recorded with Simon Scott of Slowdive on drums, another artist with more than just a passing interest in the landscape and environment.[24] More post-pop than post-rock, this album contained occasional echoes of New Order, Bark Psychosis, Talk Talk and 1990s shoegaze, as well as the provincial indie-pop of Sarah Records' bands Blueboy and Brighter, fragile guitar missives underpinned by glitchy, electronic percussion. This represented perhaps their most accessible set of songs to date, the eponymous title track featuring a distorted vocal intoning, 'Your voice will always follow me': in the accompanying video by E. L. Heath, a suburban home is viewed through grubby windows, dead flies gathering on the windowsill. There is more than just a hint of 'Ceremony' to the song, albeit written for a dead grandparent in the Midlands rather than the passing of Ian Curtis. Their next album, 2011's *Weathering*, was *The Sunday Times*' record of the week, featuring collaborations with Richard Adams from Hood and Stephen Jones (Babybird). This album riffed on themes of loss and memory, the cover a faded polaroid: the opening track 'People Say this Place is Slowly Dying' is a seven-minute track where the vocal emerges slowly from degraded washes of sound, while 'The Village Is Asleep' captures the sense of ennui associated with living on the outskirts ('I've got to leave this dead place behind').

Ben suggests on *Weathering* he was 'beginning to reflect on some of the changes I was noticing in both my immediate surroundings in the village in which I grew up and other areas nearby, the way the working-class had been left to rot since the destruction of industry during Thatcher's '80s, and how ancient institutions still remained throughout it all'. The lyrics of title track 'Weathering' encapsulate this:

> And still they stand
> The churches of England
> Still they stand
> Boarded up high-rise
> And supermarkets
> They all seem so far away
> All messed up and nowhere to go.

Despite strong reviews, after *Weathering* and an EP for French label Monopsone, Ben and Rob left Wheaton Aston, and epic45 took a seven-year hiatus. But on returning, a trio of albums – *Through Broken Summer* (2018), *We Were Never Here* (2020) and *Cropping the Aftermath* (2020) – saw epic45 sharpen their vision, distilling the essence of English suburbia with new clarity.

We Were Never Here marked another new departure for the band, including a forty-eight-page photobook the documenting familiar semi-rural non-places: phone masts, pylons, utility buildings, building sites for new red-brick estates, fly tips and bypasses. The cover – a row of blue-doored garages beneath council maisonettes – recalls the mundane provincialism of George Shaw, an artist feted for his depiction of a suburbia always devoid of people, but rich with the traces of human occupation, and oddly romantic. Shaw's paintings of his native Coventry and the Tile Hill estate where he grew are rendered in a photorealistic fashion, but using Humbrol paints of the type teenagers used to paint model planes with, hinting at a psychogeography of youth where the world beyond the estate seemed impossibly vast, the spaces surrounding the home a rich repository of personal memory.

Ben Holton suggests he 'found a kinship' with the works of George Shaw, his paintings mirroring the eye for mundane detail that runs through epic45's work. One only needs to compare the poster for epic45's 2023 tour (with Richard Adam's The Declining Winter) with George Shaw's painting *The Blossomiest Blossom* (from his 'Scenes from the Passion' series) to sense a coalescence around questions of the mundane (Figure 5.6). As Ian Waites argues, this painting

Figure 5.6 epic45 tour flyer (courtesy: Rob Glover) and poster-advert for George Shaw's 'What I did this Summer' exhibition (courtesy: Ikon Gallery).

embodies the postwar idealism of council estates, a cherry tree gloriously bursting into full bloom among 'a daisy-covered piece of communal lawn by an end-of-row house'. Waites continues by suggesting *The Blossomiest Blossom* reflects 'the remembering of a demi-paradise: a happy childhood place, familiar, decent, quiet'. But other paintings from the *Scenes from the Passion* series present a very different suburbia, one caught in 'a constantly shifting limbo' between happy childhood memories of the estate and the stigma piled upon these edgelands today. Suburbia is remembered as both familiar and safe, but it's also now graffiti-strewn, littered and abandoned.[25]

Caught between wealthy villages and the gentrifying city centre, the edgelands are then changing. As geographer Dan Dorling writes:

> England's suburbs are slowly dying and changing. The elderly are now dying faster than before. Cuts to meals-on-wheels, social worker visits, day centres, bus routes, post offices and many other suburban staples have hastened their exit. The few elderly people in densely packed terraces living nearer the centre of cities can more easily look out for each other. In old age it is in the suburbs, far away from your children (if you had children), that you now much more often live and die alone in England.[26]

These are the same the suburbs where, according to Dorling, people hold onto the dream that if they work hard enough, they will own their home, do alright and have a decent life. Their political convictions are informed by the fact that this has not always transpired: it is in suburban England that life expectancy has fallen fastest, where crime has increased most rapidly, and where transport services have often been cut. So despite promising the best of town and country, suburbia is a space where residents often feel resentment, anger, fear and loathing. Leave voters constituted the majority in the suburbs, and as Dorling notes, especially those outer suburbs which once promised so much.

This ambivalence is something that epic45's music effortlessly captures, their 2023 album – *You'll Only See Us When the Light Has Gone* – offering perhaps their final word on the schizophrenia of suburbia, a space both familiar and strange. As Ben explains, the COVID-19 lockdown was an opportunity to scrutinize his surroundings, to recalibrate and slow down the pace of life. He luxuriated in an edgelands of 'fastidious neat green lawns and kaleidoscopic herbaceous borders, weather-worn sofas and broken trampolines'.[27] The result was an album that's less subtle, and, at times, more obviously political than their previous output, a comment on the Brexit heartlands. As Ben tells me:

> It feels like the end of something, an underlining of an approach we've most probably, on a personal scale, 'perfected' by now. It's very recognisable as epic45 but we've cut loose a little more and brought in both elements of our live performances and a naked love of pop. The title is indeed a slightly tongue in cheek nod to the way we've been toiling away in the margins for so long, ploughing our lonely furrow. But there's a wider sense to it also. Lyrically, I'm looking outwards more, at the people I see around me, hopefully in a non-judgemental way, just noticing and reflecting. I'm looking at people who maybe feel abandoned and forgotten, lectured at by politicians, pundits and celebrities: some are sadly leaning into age-old prejudices, stoked by the aforementioned media channels, trying to find meaning and something to believe in.

The cover depicts the Highfields Estate on edge of Stafford, with Cannock Chase rising behind it in the distance. Is this one of England's 'left behind spaces', perhaps? The album is certainly about those who live in a forgotten and largely ignored part of Britain, but Ben argues theirs is an empathetic engagement with the edgelands, a search for a way to survive that does not end in the 'falling motion' of blaming others for the ills of Middle England.

The album's opening track – 'New Town Faded' – alludes to the lives of those who feel dead inside, who have lost hope in the planners' utopia, and in some cases have turned to the right-wing politics of suspicion and resentment. 'Be Nowhere' contains some of epic45's most direct lyrics to date, accusing suburbia of 'Eating you alive, fucking with your mind': its image of a crestfallen woman doing the washing up, marigold gloves on, dolefully looking from the window of her flat onto distant countryside, is a powerful summation of life in a 'fallen English town'. On the elegiac 'Floodplains' we hear of those living in sub-standard houses built on zones of flood risk, who saved for the future but found themselves trapped in debt, unable to move and burdened with the wrong education, the wrong identity. The last track – fittingly entitled 'Finality' – builds to an angry crescendo, the vocals near-screaming in despair:

> I have wasted the small hours
> Waiting for answers
> Too much time has gone
> Waiting for the morning light to save me
> No way to go back
> Waiting for tomorrow
> We're leaving home
> Across country
> To another place
> To another town
> We'll stay on the outside
> And watch life from the inside.

Whether or not their thirteenth album is epic45's last, it seems a suitable epitaph for the English edgelands, a realm of uneasy hybridity haunted by the thwarted hopes of youth.

Further Reading

Edensor, Tim (2005). 'Waste matter – the debris of industrial ruins and the disordering of the material world: The materialities of industrial ruins'. *Journal of Material Culture,* 10(3), 311–332.

Farley, Paul, & Roberts, Michael (2011). *Edgelands: Journeys into England's true wilderness.* Jonathan Cape.

Gandy, Matt (2023). *Natura Urbana: Ecological constellations in urban space.* MIT Press.

Grindrod, John (2017). *Outskirts: Living life on the edge of the Green Belt.* Hachette UK.

Houghton, Frank, & Houghton, Sharon (2015). 'Therapeutic micro-environments in the Edgelands: A thematic analysis of Richard Mabey's The Unofficial Countryside'. *Social Science & Medicine*, 133, 280–286.

Mabey, Richard (1973). *The Unofficial Countryside.* Collins.

Mabey, Richard (2024). *Accidental Garden: Gardens, wilderness and the space in between.* Profile Books.

Smith, Joe (2017). *The New Nature Writing: Rethinking the literature of place.* Bloomsbury.

Stoetzer, Bettina (2022). *Ruderal City: Ecologies of migration, race, and urban nature in Berlin.* Duke University Press.

Waites, Ian (2013). '"Places where I forgot things": Memory, identity, and the British Council Estate in the paintings of George Shaw'. *Cultural Politics*, 9(3), 357–370.

6

Weird Walking and the Prehistoric Past

It's late March 2024 and England is waking from winter slumber, the continuing wet weather breaking for a rare day of spring-like sunshine. In the heart of London, the parks are beginning to buzz with life, and Sunday afternoon tourists meander towards Buckingham Palace. But just off The Mall, underneath Carlton Park Terrace, a different crowd is gathering in the somnambulant gloom of the Institute of Contemporary Arts, a venue that down the years has put on a rich diet of experimental art and music. The attendees on this Sunday afternoon are diverse, mainly in their thirties and forties, appearing left-leaning, queer and metropolitan in the main, but interspersed with older attendees exuding more hippyish and scholarly vibes. A few minutes before two o'clock, the crowd lolling around the venue's café and bookshop are ushered through to the main auditorium, many heading straight for the floor cushions arranged around the stage in anticipation of the event to follow.

The event, headlined 'The Secret Life of Stones', is described as aiming to stimulate a discussion about 'stone consciousness': the compere begins, with no sense of irony, by introducing her favourite stone. Then the lights are dimmed, and a short film, *The Stones of Kernow* by Matthew Shaw, is shown. Eschewing any narrative, this is black and white footage of stone circles, dolmens and megaliths in the South-West of England: Men Gurta, Boscawen-Un, Dry Tree Menhir, Castle an Dinas, Carn Euny. The camera pans slowly around the stones, taking in their texture and form, at other times zooming beyond the circle to emphasize the placement of the stones in relation to the wider landscape. The music that accompanies this mainly consists of slow, beatless drones, seemingly emphasizing the longevity of the stones in the landscape, gentle synth melodies occasionally overlaying these, adding hints of mysticism and grandeur. After twenty-three minutes, the camera pans skywards, and the credits roll. The crowd claps loudly, with enthusiasm.

This is Stone Club, or at least one of a series of regular events held at venues the length and breadth of the UK, organized by Cornish-based artist Lally MacBeth and her partner, musician Matt Shaw, since 2021. These encourage an inter-disciplinary discussion between archaeologists, historians, filmmakers, writers and musicians about Neolithic landscapes, this particular event taking in discussions of new books on landscape (in this case, academic Noreen Masud in conversation with Justin Hopper), a film by Jeremy Deller about the outsider artist Bruce Lacy, and an enthusiastic sing-along led by Gwenno. Her presence is noteworthy given her Mercury-nominated Cornish-language album *Tresor* (2022) was preceded by a single – 'Men an Toll' – referring to the holed stone at Land's End understood to be an ancient fertility symbol (Figure 6.1). Other Stone Club events have occurred at folk and music festivals (End of the Road, Glastonbury, The Moseley Folk Festival) and across an esoteric range of bookshops, art cinemas, clubs, gardens and churches. Musical contributors have included a diverse range of performers from the world of contemporary folk, psychedelia and freak-out experimentation, including Bill Ryder Jones, Gruff Rhys and Richard Norris (aka Order of the Twelve).

The popularity of these events has overlapped with the emergence of a range of psychogeographically infused fanzines and books – including *Boulderdash* (stones, drones and noise), *The Wulver's Stane, Hwaet!* and *Grimoire Silvanus* –

Figure 6.1 Mên an Tol, Land's End. Courtesy: PickPic royalty-free images.

all riffing on the psychic charge of prehistoric sites. Perhaps most prominent is *Weird Walk*, which spawned an accompanying 2023 book offering a suggested seasonal itinerary of visits to ancient and pagan sites of Britain. Proposing a psychogeographical engagement with the landscape that bypasses nostalgia to seek re-enchantment, *Weird Walk* was started by Alex Hornsby, James Nicholls and Owen Tromans, three friends who bonded when walking the Wessex Ridgeway (which passes through Avebury).[1] Like Stone Club, *Weird Walk* promotes shared experiences of countryside rambling, embracing a mixture of magic, myth and music: *Weird Walk* has notably spawned a music zine spin-off, *Peasant*, which focuses on the interest in folklore across genres including dark metal, dance and electronica.[2]

All this hints at a current enthusiasm for the landscapes of English prehistory at odds with classical archaeological and anthropological attempts to excavate the 'true meaning' of these stones, being playful, creative and just a little provocative. Stone Club events are not dry academic discussions of theories of stone alignment or presentations of artefactual evidence but celebrations of the numinous qualities of stones that deny any specialized knowledge or insight is necessary to enjoy visiting these ancient sites. Notably, no one guffaws when a contributor to the ICA Stone Club mistakenly locates Stonehenge in Somerset (it is in Wiltshire): it simply doesn't matter. The ethos of Stone Club is inclusive and collaborative: as their website proclaims, 'the stones are for everyone' and 'there are no rules'. In short, it doesn't suggest there is a singular 'correct' interpretation of stone landscapes – which have been variously posited to be timepieces, altars or graveyards – but suggests these can be integrated into any number of personal searches for meaning and identity.

The inclusive ethos of Stone Club also rejects any attempt to integrate British prehistory within nationalist discourses proclaiming an essentialist basis to Englishness, not least those who proclaim whiteness as a 'primordial ethnicity' or suggest that non-white populations cannot understand, or appreciate, prehistoric sites. As sociologist (and sometime Stone Club contributor) Ben Pitcher argues, it's just as valid to challenge racism and xenophobia by considering who is welcome to experience and enjoy the stones as it is to reflect on, for example, the over-policing of black bodies in the inner city, or the naming of civic buildings after seventeenth-century slave traders.[3] In this respect, the formation of Stone Club seems a reaction to attempts to proclaim Neolithic monuments as inherently 'British', including explicit references to them in right-wing discourse, such as Jacob Rees-Mogg's impassioned 2019 election video filmed at Stanton

Drew stone circle, in his North East Somerset constituency, which argued for the importance of getting 'Brexit done'.[4]

Stonehenge and Avebury are perhaps the best-known archaeological sites in the country, and as such are particularly prone to being caught up in projects of national identity-making. But there are more than a thousand known stone circles in the UK, and innumerable henges, ditches and standing stones proclaiming Neolithic origins. While generations of prehistorians and archaeologists have provided gazetteers of these, it's Julian Cope who is often credited with making the stones accessible to wider publics, eclipsing learned authorities like Aubrey Burl or William Stukeley. Cope's pre-millennial *Modern Antiquarian* guide appeared when his top forty hits had begun to dry up and he was re-styling himself as the arch-drude, a savant with insight into pagan and the occult obtained through devotion to unhinged shamanic krautrock and Norse mythology. Splitting with major record labels, his newly formed Head Heritage label allowed him to indulge his passion for meditational heavy metal and exploration of prehistoric Britain. His 1998 guide to 300 of Britain's stone circles and ancient monuments – tellingly dedicated to the culturally dispossessed of Britain, be they 'white, black or green' – found a receptive audience, followed in 2004 by *The Megalithic European*. Both fuelled heated discussion on the Modern Antiquarian website, which came to be a major online platform for the enthusiastic sharing of stone photos, arcane theories and fieldnotes.

The co-founder of Stone Club, Matt Shaw, recalls Cope's influence on his own musical experiments in field recording: 'It came at the right time for me age-wise as I was very open to these ideas . . . Musically, I have always loved Julian's songs but it was his more ambient and experimental work that had a literal influence on my music'. Shaw also credits Cope's book and website with bringing him into contact with other stone enthusiasts. As he recalls, 'There was this combination of people that were really interested in music and antiquarian sites, and all the things that connected around that which could be incredibly political too: it could be road protesters, or it could be people that just really liked Amon Düül and Kraftwerk'. Among these was Mark Williamson, who records as Spaceship, regularly posts stone reports to his YouTube channel and has released numerous songs recorded within megalithic structures. Williamson recalls meeting Matt Shaw through their shared enthusiasm for stones:

> We met kind of through the Julian Cope phase when he was writing these books about stones. In that kind of first wave of it there were a lot of people

> really enthusiastic for visiting these ancient sites, connecting with them not in a nationalistic way, not in a 'this is the English culture' way, but just because these are ancient sites. But what happened over time, as happens with these things, is everyone visited all the big sites, everyone's gone to Avebury, Castlerigg and Wayland's Smithy. And the trainspotterism started to infiltrate. So you then get people saying 'I went to the round barrow on something hill' and then someone was 'That's not a round barrow it's a disc barrow' and all the fun went out of it.

This hints at some of the tensions around stone enthusiasm, and partly explains the impetus to the formation of Stone Club, which emerged in 2021 as Shaw realized that, after lockdown, he wanted to travel beyond his immediate locale (at the time, Dorset), and visit stones, not in an attempt to proclaim any particular expertise, but to participate in something *shared*:

> It was in response to a world that felt increasingly binary and divided. Leaver or Remainer, pro-vaccination or anti-vaxer, climate change activist or denier – on top of which came the first war in Europe in my lifetime [*in the Ukraine*]. Why was there so much disagreement and such strength of feeling at each end of the spectrum? The reasons are many and complex, of course. But what if we could all agree on one thing? What if we could agree that a stone circle is an amazing thing? The reasons why we could bond over stones would be many – from the academic to the spiritual. Archaeologists, geologists, poets, musicians, writers, dowsers, those of any faith, ramblers, right-to-roamers, dog walkers, and anyone out walking just to feel closer to being a part of Nature would all be equally welcome. If we could agree on liking, loving or at least having an interest in one shared thing, then surely there would be . . . the beginning of a new (ancient) inclusive space for us all?[5]

As he quickly found, this idea resonated with others, and Stone Club was born.

In an interview with Justin P. Hopper, Shaw recalls asking people to send in their pictures of walks to stone circles, not asking for details of their geology or morphology but more mundane details of who they travelled with, and how. He found people would send photographs of themselves, 'say age 8 or 10 and then them in the exact same spot aged like 30-something'. He confesses 'there was something really magical about looking at those two photographs side-by-side in one person's life, as the stones look almost exactly the same, the landscape often looks almost exactly the same . . . so you could stand in that place and you could remember a moment in your life that that perhaps had evaporated'. Lally also recalls one woman sending 'a picture of a man with his head through a stone, asking us if we knew where it was. We told her it was Men-an-Tol and she explained it was her father, who had since died. She came on holiday and visited the site with her boyfriend, who proposed to her at Men-an-Tol. That really touched us'. Matt concludes 'it's quite magical to be able to unlock some of those feelings, emotions, sensations'.[6]

This interest in capturing these sensations through music began for Shaw when he was growing up in Cheshire, as he tells me:

> My first recordings were made in Sandbach and Cheshire where I grew up. There were two things I Intended to do: one was to capture and preserve the everyday sounds I knew so well from home and my locality. The particular birds in the garden, sounds that only collided or existed because of their exact location. This wasn't because I thought them special in any way really, it was more personal. They were sounds that surrounded my life with such familiarity but I was aware from a young age that the combination of factors creating that audio environment was slowly changing and one day would be very different. So I wished to capture the sound like a photograph to preserve the memory. The other thing was recording people, their memories and stories. This started with family and preserving stories that might otherwise have been misremembered or forgotten.

For Shaw, the first Stones that fascinated him were the ancient (actually ninth-century Saxon) crosses in Sandbach. As he explains 'they were just always there, and very normal'. But as he was growing up, he came to realize that not every town had these: 'I think that was that was something that was really central, you know, to my earliest memories, that his place had these and perhaps made that town different than some of the ones surrounding it'.

Shaw subsequently played in a variety of punk, riot grrl and post-rock outfits, before a series of releases as Tex La Homa (a skewed take on Americana). But his interest in the prehistory of landscape persisted, and when he relocated to Dorset in the 1990s he began to produce a series of experimental and primordial drone releases inspired by the landscapes of the South-West, partly drawing on the ambient work of Brian Eno: *Metherell* (Sonic Oyster, 2010), *Sandbanks* (Quiet World, 2012), *Seatown* (Siren Wire, 2012), and *Fowey* (self-released, 2013). *Pathways of Ancient England* (self-released, 2022) and *The Still Silence at the Centre* (2024) further demonstrated his interest in journeying along numinous pathways, the former album being recorded on the ancient routeways of Britain, starting in Christchurch on three hills that aligned before travelling to the Welsh mountains and numerous ancient sites on journeys from Land's End to Anglesey. These in situ ambient recordings were the product of his desire to connect more deeply with the natural world, and, in his words, to seize the 'atmosphere of place', something also reflected in his prose, poetry and photography collection *Atmosphere of Mona* (2020) published alongside an accompanying soundtrack / audiobook. In an interview with Gary Cook in *The Ecologist* (2017), he describes his field music 'as primitive photography', seeing it as a way of recording a sense of place that can 'never be repeated'. As he recounts, 'Where I position the recorder will catch the sound of the breeze, waves lapping the shore or fishermen unloading the catch from their boats and, if it were recorded again, it would inevitably be slightly different. In that way, it is naturally nostalgic, creating its own sonic footprint, constantly changing, evolving'.[7]

In a revealing 2024 essay for *Caught by The River*, Shaw gives further hints as to how this interest in place and landscape inevitably led him back to the stones. On moving to Dorset from Cheshire in the 1990s he admits he didn't know the county very well outside of the major centres and the recording studio in Bournemouth he worked in. But he bought a car and began to engage with the histories and geographies of Dorset by visiting the sites listed in Cope's *Modern Antiquarian,* the first being Nine Stones near Winterborne Abbas, in 2004:

> The Nine Stones, also known locally as the Devil's Nine Stones or as The Nine Maidens, dates from the late Neolithic or early Bronze Age period. It is a small circle, around nine metres by eight meters in diameter. There is a wonderful illustration of the stones from 1723 by William Stukeley titled 'A Celtic Temple at Winterburn' [*sic*] that depicts the circle with what is now the A35 running past. The landscape illustrated is much more open than today, the circle more prominent, but it's still instantly recognisable. The folklore surrounding the stones includes a story of how the stones are the Devil, his wife and their children, another story claiming the stones are children turned to stone for playing on the Sabbath.[8]

The fact the Nine Stones are alongside the A35 now makes the site difficult to visit or photograph (Figure 6.2). But for Shaw this was perhaps part of the appeal: he confesses that 'trying to find the lesser known or harder to find places became like a game', his Ordnance Survey map flapping around in the wind and getting soaked as he kept a beady eye out for farmers who might accuse him of trespass. Each visit to the stones was also accompanied by background reading: he tells me he consumed books on 'folklore, history, prehistory, psychogeography, the Occult, travel and occasionally the odd novel'.

Figure 6.2 Nine Stones, Dorset in Stukeley's 1723 illustration and in the contemporary. Courtesy: Ethan Dayle Whyte / Creative Commons.

Shaw's visit to the Nine Stones, alongside a range of sites in Cornwall, provides the basis of his most complete musical engagement with the prehistorical sites of England to date, *A Journey to The Stones* (2024). On this album (Figure 6.3), field recordings and samples were fastidiously assembled by Shaw within the granite walls of his home in Penryn, near Falmouth, where he now lives with Lally MacBeth, a Cornish native. More cohesive than his previous releases, each

Figure 6.3 Matt Shaw's *A Journey to the Stones.* Courtesy: the artist.

track refers to an individual stone and its unique resonance, the album a journey best listened to sequentially. It begins by the A35 in Dorset at the 'Nine Maidens', a track that has percussive bleeps and clicks giving the track momentum while, in the background, a steady drone increases in volume and pitch before steadily fading, followed by 'Castle An Dinas' (named for an Iron Age hillfort near St Columb in Cornwall) and the languid 'Hellstone' whose gentle pastoralism is undercut by faint voices that seem to echo around the Neolithic dolmen. In contrast, 'Men Gurta', Cornwall's largest monolith at nearly five metres in height, is represented through shrill whistling drones, the accompanying YouTube video showing the monolith standing as exposed as the wind turbines behind it, which turn slowly. Though this standing stone gives the impression of permanence and stability, in contrast to the turbines that are newcomers to the landscape, it actually fell in 1945 and was only righted again by the Ministry of Public Building and Works eleven years later. Many other standing stones, cairns and dolmens have of course not fared so well, with regular incidents of graffiti, vandalism and fire-lighting having drawn attention to the need to police 'heritage crimes', providing a justification for reducing public access to the stones.

Music of the Megaliths

Stone Club has clearly tapped into the popular resurgence of interest in prehistoric sites, with Matthew Shaw's drift from a broad interest in environmental sound to one registering the lives of stone circles indicative of a twenty-first-century coalescence of trends in tourism, archaeology and popular antiquarianism that have rejuvenated interest in these mysterious sites. Indeed, alongside Matthew Shaw and Mark 'Spaceship' Williamson, recent years have witnessed a slew of other musical acts engaging with stones with an enthusiasm not witnessed since the early 1980s when bands including Hawkwind and Ozric Tentacles would regularly play to the disparate 'tribes' of travellers, heads and peaceniks who gathered at the Stonehenge Free Festivals to see in the summer solstice.

Acquired for the nation in 1918 and taken under the management of the National Trust in the 1930s, Stonehenge has long been a contested site, inspiring different forms of appreciation because of its ambivalent symbolism. On the one hand, Stonehenge signifies the potent strength and endurance of the human endeavours of the past, yet it remains vulnerable to insensitive development and modern encroachment. The latter concerns certainly came to the fore in the 1970s as conservationists started to raise the alarm about the impact of the Free Festivals on the site, with the stones roped off from 1977 and the police preventing visitors from parking on the field adjacent to the stones. These exclusionary actions undoubtedly reflected underlying moral anxieties about the exuberant excess of the Free Festivals, with David Matless suggesting Stonehenge became caught between two visions of pastoral Englishness, one insisting that it needed protection for the nation, the other that it represented the essence of 'free Albion' and needed to be accessible to all.[9] This culture clash reached its apogee in the 1985 'Battle of the Beanfield' when a Peace Convoy of 600 or so New Age travellers was prevented from reaching Stonehenge, chased through the fields by up to 1,300 riot police attempting to maintain a four-mile exclusion zone around the stones. Viewed retrospectively, photos of police wielding truncheons against women and children suggest that much of the media's depiction of the travellers as violent and aggressive towards the authorities was the product of an orchestrated attempt to depict them as a modern-day 'folk devil'. The subsequent

Public Order Act 1986 was a classic piece of moral panic legislation that, in part, used the protection of Stonehenge as a justification for a suite of powers imposing limits on public assembly, as Penny English writes:

> Authority was used to an extraordinary extent to exclude particular people and activities from the monument, resulting in the paradox of barricades separating a publicly owned national monument from its citizens. It was precisely because it has a symbolic national value that appropriation of the site could be portrayed as an attack on the values and identity of society itself, thus justifying the measures taken to protect it from socially unacceptable activities. Conflict at the site was not just about the use of the monument but reflected wider issues, reinforcing the strong links between cultural heritage and its political and social context.[10]

In the twenty-first century, debates about public access to, and enjoyment of, Stonehenge continue to simmer, not least because of the planned tunnelling of the A303 under the site, approved in 2023 to the dismay of many archaeologists who feel it threatens the integrity of the site.[11]

Today, some clearly still hark back to the days when Stonehenge's Free Festivals constituted a vibrant and unbridled expression of a counterculture that encompassed neo-paganism, environmentalism and musical experimentation, one that lived on in the parties organized by Spiral Tribe in the early 1990s in nearby Longstock – at least until the *Criminal Justice and Public Order Act (1994)* curtailed rural raves and the playing of music 'characterised by the emission of a series of repetitive beats'. Several recent musical tributes to Stonehenge hence reference the A303 as the route that took Londoners down to Wiltshire and Hampshire for these gatherings. For example, ex-Ozrics performer Merv, later of Eat Static, released the nine-minute psytrance of 'A303 (Stonehenge Turn Off Mix)' as Kashmir Black in 2021, while OSC's (2023) 'Stonehenge (Ancestral Landscape remix)' is a classic slab of hypnotic deep techno that also hints back at the days before the *Criminal Justice Act*. Others seemingly embrace Julian Cope's no-nonsense approach to the antiquarian, with Dorchester-based Ætheling's *West Dorset* album combining stoner rock and primitive kosmiche on tracks including 'A303' and 'The Hellstone' (the latter referring to a long barrow near Portersham on the ancient South Dorset Ridgeway)

Luke Murray's Stonecirclesampler project, on the other hand, offers a jungle-infused exploration of the role of standing stones in sonic rituals past and present, suggesting 'Avebury was the original club', and rhetorically asking, 'if you played back the last twenty years of club music into Stanton Drew . . . what would they play back to you?' His twenty-five-minute 'Megalithic Grime Radio Documentary' combines soundbites from Cope's *Modern Antiquarian* with grimy drum and bass to create a soundscape hinting, in his words, at both the 'distant past and future-to-come all at once', while *Deep Dream Derbyshire Gloom* (on Brachliegen Tapes) refers to Peter's Stone and Bleaklow in the Peak District, skittering slices of dark techno hinting at 1990s raves and nefarious congregations at stone circles, events whose memory now infuses the topography of the Peaks. A limited-edition cassette on Middlesbrough's Industrial Coast label, *Save The Stones!*, appeared in 2021, followed by a sixty-minute dub version *Erode The Stones!*, the latter suggesting the industrial techno and dubstep movement began in the shadows of the stones as much as in the pre-gentrified streets of London, Manchester or Bristol. In 2024, a limited ten-copy lathe-cut single *Avebury Ambience* brought Murray's Neolithic driftwork to a close, leaving him to focus on his Superior London Pulp project and hallucinatory hauntological excursions elsewhere in the country.

Before adopting his Stonecirclesampler alias, Murray worked at Andy Votel's Finders Keepers record label in Manchester, alongside Sean Canty, one half of Demdike Stare, a band whose dark ambient music draws on the folk histories and 'witch' cultures associated with landscape of the Pennines, and whose time living in Burnley led them to explore the standing stones of the moors. Demdike Stare were often cited as de facto leaders of the 'witch house' scene that rose to prominence in 2009–10, something they encouraged by regularly playing to a backdrop of esoteric films including the 1922 Swedish documentary film *Haxan*, a documentary on English witchcraft. Their interest in collisions of past and present, use of library music and film scores, and occasional nods towards the Radiophonica of Delia Derbyshire et al suggests much in common with the hauntology of Ghost Box, but musically their uses of drones, drum and bass and industrial beats suggests a fusing of ancient and modern that's more about repositioning the dead past in the present than working through ideas of lost futures.

In his remarkable book *Albion's Eco-eerie*, an exploration into culture's obsession with its Other in the dark times of environmental and social crisis, Phil Smith argues that ancient stones are one of the privileged places where echoes of English 'lore and practice' are at their most intense.[12] The idea that stones act as a repository of repressed memories of the occult and demonic, and have borne witness to a secret history, hence feeds into a musical approach to stone circles and megaliths which engages with the themes of *folk horror*. Only one of the original 'unholy trinity' of folk horror films – *The Wicker Man* (1973)

– featured stones in any significant way, and even then the stone circle in the fire-leaping initiation scene was merely a movie prop. But the notion that stones are conduits for supernatural or cosmic energies has nonetheless been a recurrent trope in British folk horror, inspiring much of the music that has subsequently explored this tradition. For example, alongside Belbury Poly's *From an Ancient Star*, numerous other musical acts have drawn inspiration from Avebury's sinister depiction in *The Children of the Stones* and the more general inclusion of stone circles in the disquieting children's television of the 1970s, including *The Changes*, *Sky*, and *Raven*.

Exemplary here is Jonathan Sharp's 'dark agrarian' The Heartwood Institute project. This draws on both his love of the music of the Radiophonic Workshop as well as memories of childhood visits to Castlerigg stone circle in his native Cumbria, the stones situated evocatively in the shadow of Helvellyn. These influences are readily apparent on his 2017 album *The Whispering Knights*, which draws directly on a young adult novel he originally read as a child, Penelope Lively's book of the same title, published in 1971. As he tells me 'I've got a fairly mixed set of inspirations, from nature fiction like Sarah Hall and Penelope Lively through to the occult, folklore, counterculture and protest . . . a real mixed bag'. *The Whispering Knights* hence offers a suitably spooky imagined soundtrack to Lively's children's novel, which is centred on the Whispering Knights dolmen burial chamber, consisting of four upright stones and a large fallen capstone, situated near the Rollright Stones in Oxfordshire (Figure 6.4). The monument obtained its name via a legend about the king and his army being petrified by a witch, the stones now imagined as the treacherous knights conspiring against the king. In Lively's novel, children living in the valley below the stones inadvertently summon the ghost of Morgan Le Fay, with the culmination of the novel seeing the 'knights' coming back to life to defeat her. Sharp's album offers a faux authentic soundtrack for a lost 1970s film version of the book, lilting pastoral interludes ('Steeple Hampden') and medieval reverie ('Clipsham Manor') interspersed with sinister and macabre stabs of *giallo* noise ('Morgan pays a visit') and a squelchy Radiophonic conclusion (The Battle), all bookended by the portentous 'Whispering Knights theme'.

Somewhat coincidentally, Gordon Chapman-Fox (who also records as Warrington-Runcorn New Town Development Plan) released the LP *The Nine Travellers* (2023) which included the track 'The Whispering Knights' alongside three other long-form modular meditations on the mythology surrounding stone circles: 'The Grey Maids', 'The Black Guardian' and 'The King's Stone'. Here, the

Figure 6.4 The Whispering Knights, Rollright Stones, Oxfordshire. Courtesy: Wikipedia Commons.

King's Stone is an outlying standing stone also situated near the Rollright Stones. The reference to the Grey Maids is more obscure, but the reference to the Black Guardian suggests this is a playful citation of the 1978 Doctor Who episode *The Stones of Blood*, one of the BBC series' rare excursions into folk horror. The working title of this was 'The Nine Maidens', the story fusing fears of druidic ritual to a far-fetched plot that identifies the stones as alien Ogri from the planet Ogros. Part of the 'Key to Time' season which saw Tom Baker's doctor fighting to defeat Valentine Dyall's memorable Black Guardian, the story was set in contemporary Cornwall, but with the Rollright Stones used for the filming of a stone circle, set designer John Stout adding artificial cross-pieces to the real stones, similar to those seen at Stonehenge. Here, Douglas McNaughton notes the serial's debt to the conventions of folk horror, exaggerated through the profusion of Celtic names (the Cailleach, Morrigu, Ceridwen and the Arthurian Vivian Fay); the folklore of Cornwall, including giants and ogres in the form of the Ogri, and the local myth of being unable to ever truly count the Rollright Stones.[13] The opening scene of a pair of campers being crushed by a stone also summons folklore about the megaliths falling on the unwary, a myth sedimented by the genuine discovery of skeletons beneath some stones, presumed to be those killed during their construction.

Contemporary experimental and hauntological music is then not innocent in its layering of fictive depictions of stone mythology onto the existing folklore of stone circles, both exploiting and adding to the eerie and disquieting reputation associated with stones in the landscape. For example, David Chatton-Barker and Ian Humberstone's (2016) *Megalithic Monuments Volume 1: Stanton Drew Stone Circle* focuses on the legends of the Somerset stone circle being the petrified remains of a ribald wedding party, the opening spoken word piece being a recounting of this legend accompanied by a music box, with the rest of the EP consisting of two scratchy, echoing suites recorded on xylophones played on top of the stones. The accompanying sixteen-page booklet includes multiple photos of the stones degraded by material found at the site, suggesting the staining of time: tellingly, the EP is among a slew of releases Folklore Tapes have donated to the Museum of Witchcraft and Magic in Boscastle, Cornwall.

Ancient atmospheres

Given no straightforward explanations exist for the form of stone circles and megaliths, it is hardly surprising they are repositories for a remarkable range of music and stories involving druidic ceremony, occult practices and restless spirits.[14] In this sense, the invocation of stone circles as sinister sites is in danger of becoming an all-too-convenient way of adding depth and mystery to musical projects. For example, Haven Avenue's self-titled (2019) debut album offers folkish strumming over field recordings made at Avebury, all breathy ethereal vocals and allusions to standing stones as the scene of grisly murders, while at the other end of the musical spectrum, black metal is full of stories of sacrificial, blood-stained, stones.

So while stones feature widely as cognitive anchoring points in invocations of folk memory and pagan presence in the landscape, much of this music appears concerned with invoking a generalized atmosphere of mystery and suspense rather than offering a response to the materiality of the landscape. In this vein, Tim Edensor complains that when making sense of stones many commentators focus simply on their appearance, reifying this into cultural representations that project meanings onto these forms: in so doing, the real substance of these sites is neglected, namely the 'textures of sounds, smells and tactility . . . that herald landscape's liveliness'. As Edensor notes, though the purpose and meaning of standing stones will always remain mysterious and inscrutable, 'it does not take a leap of the imagination to consider that the potency of these sites' depends upon 'the play of shadow and light across the . . . stones and the wider landscape'.[15] Here, Edensor notes that many stones stand in exposed locations, effectively connecting earth and sky: visiting the stones one becomes especially aware of the position of the sun as it moves during the day, casting shadows and leading the eye in different directions.

As such, Edensor argues it is difficult to summon the uncanny elements of place without taking account of embodied experience, suggesting an important role for music in evoking the phenomenological encounter with stones. This suggests a need for creative attempts at summoning the *feeling* of visiting Neolithic sites. Exemplary here is Kemper Norton's *Carn,* the first half of which focuses on his encounter with the Carn Euny fogou, an astronomically aligned Iron Age chamber in West Penwith, Cornwall. Much of this was recorded within the mysterious fogou itself, whose ancient function was thought to be ceremonial but remains obscure. Norton's album begins with '22/8/99' marking the date of his visit near the turn of the millennium, at the time of the solar eclipse:

> Me and a friend stayed down there for the eclipse, we camped there the night before and it was a kind of unusual experience. It's just a very special place really, and I recorded down there. I brought down a Casio keyboard and played with that down there. I don't really believe in ghosts or anything, but there's definitely a presence there and I didn't want to be taking the piss. I was trying to do it in a way that was respectful as well. I'm not sure I'd call it a supernatural experience, but I definitely experienced something that I don't want to talk too much about, that was a bit sort of strange. I'm not generally a ghost-spotting kind of person, but it had a very big impact.[16]

The recordings that Norton made in situ were woven into a studio-compiled sound collage which mashes computer glitches, harmonium drones, snatches of obscure folk songs, snippets of dialogue and even traces of rave, to suggest the site is alive with the spirit of the past, as if the fogou were a stone tape. Along the way, *Carn* alludes to the restless ghost of Jon Tregeagle, who was charged with carrying sand up to the Carn from the nearby Vellandreath Valley, suggesting local folklore shaped Norton's own expectations about the atmosphere he would encounter there.

Some contemporary electronic performers have though gone further in their sonification of the stones by embracing the archaeo-acoustic theory that stone circles once served as giant amplifiers, cutting out sound outside the circle but making it easier for those inside to hear and participate in music-making. This is manifest in an interest in the resonant frequency of the stones, and while few go as far as the University of Maine researchers who have at a full-size replica of

Stonehenge in the US in which they use reconstructions of Neolithic instruments, several artists have used in situ field recording to pick up on the 'frequency' of the stones.[17] For example, Matt J. Sanders (as Assembled Minds) released 'Refracting Blue Stone Light Formation' on 2018's *Psychogeographic Audio Explorations Along the A303, Volume One,* the track recorded in situ and in real time at Stonehenge on a primitive Teenage Engineering OP-1 synth. The track captures Matt's emotional response to the surroundings, weather and history, a meditative response to his surroundings in which the presence of the stones is a direct influence. Somewhat relatedly, Bioni Samp's (2013) 'Electronic Journey To Avebury' involved him using a 'Binaural Beeframe EMF Detector', 'Hive Tool Resonance Detector' and video camera to register the soundscape. This involved readings of the electromagnetic radiation and transmitted resonance at Avebury, mixed with homemade instruments including a 'hive synthesizer' producing frequencies akin to those of bees via modular oscillators. The result is a strange, buzzing, sixteen-minute track that at times sounds like an engine throb, and implies the stones are a strange hybrid of the organic and non-organic.

Samp's semi-scientific approach is partially mirrored in the work of Ben Powell. As a field support technician from Bangor University, his day job involves supporting field researchers using sensors to remotely sense changes in the extent of Arctic ice, ocean currents and surface temperatures. But alongside this, Powell has made more than a dozen ambient releases since the late 1990s, firstly as Mank, and latterly as Llyn Y Cwm. These releases have aimed to capture the atmosphere of environmental sites including sub-glacial lakes and aquatic biomes, but increasingly stone circles, many in his native Wales. Of necessity, many of his releases have been made when at sea on a Research Vessel, using a compact laptop in lieu of a modular synth, and virtual studio technology plugins. With some guitar overdubs, inspired by Steve Hillage and others, these tools enable him to create dark ambient soundscapes:

> I grew up listening to Iron Maiden, but in the 90s I discovered electronic music via Godflesh through to Warp records. In 1996, I bought a cassette 4-track, drum machine and a guitar FX unit, I would record guitar through a huge reverb, then turn the cassette over and add more reverb in reverse, then repeat – I ended up with something that sounds similar to what I am making today. I am totally self-taught and tend to use effects over studio 'know-how'. I discovered the music of Steve Reich via The Orb, and *Music for 18 Musicians* remains my favourite piece of music: I find the evolving patterns and rhythms endlessly fascinating.

Released in 2023, his album *Megaliths* is a 'sonic voyage that mirrors the sacred spaces marked by stoic pillars of rock', each track representing a specific prehistoric site. Guided by Cope's *The Modern Antiquarian* and Aubrey Burl's classic *Guide to the Stone Circles of Britain*, Powell – as Llyn Y Cwn – captures the remote bleakness of these sites, portraying them as structures where, in his words, 'the veil between our world and the eternal realm becomes tantalisingly thin'.

Based on years of exploration, Ben has collected field recordings capturing the 'haunting whispers of wind and the solemn caws of crows', enveloping the listener in an atmosphere that transcends the ordinary. This search for transcendence is one of the reasons for Ben's interest in stones, yet, as with Matt Shaw, this combines with the more earthly pleasures of 'stone bagging':

> Aside from music, searching for ancient megaliths, stone circles and burial sites is my geeky hobby. It is the thrill of the chase, a goal at the end of a damp walk. I think it started when I moved to Anglesey about 25 years ago . . . the island is full of ancient sites. I bought the Aubrey Burl guidebook and started searching using OS maps. Julian Cope's *Modern Antiquarian* is another well-turned book I own. I spend a lot of my spare time traveling the country in my campervan visiting these sites, I love the mystery, the fact that no one really knows why they were made.

Megaliths invites the listener to experience stones that are 'strategically positioned to occupy the centre of the world', acting as microcosms of the surrounding landscape, something reflected in the eighty-eight pages of photos in the booklet that accompanies the release, high contrast black and white images emphasizing site and situation (Figure 6.5). As Powell explains, the photography and the music are interlinked, and if he sees or hears something he will 'set the recorder going, step away and take some photographs' which become source material for his compositions. There appears a form of synaesthesia in play here, with Powell suggesting he processes photos and sounds in a similar fashion, 'reducing the colours and vibrance, emphasizing the minor details and increasing contrast'.

Megaliths consists of ten tracks invoking named standing stones, cairns and barrows across Cornwall, Devon and Wiltshire as well as The Isle of Arran (Machrie Moor), Ireland (Cashelkeety) and Wales. In each case, Powell attempts to capture the situated experience of being in the presence of the stones:

Figure 6.5 Bryn Cader Faner. Courtesy: Ben Powell.

> The album is based on field recordings made at various megalithic sites I started visiting these sites about 25 years ago. I think I started making recordings to act as an aural memory book. I can play the recordings back on my headphones and I'm instantly transported back to that time and place: I find it more immersive than video. I like to take small details and stretch them out, they can then either be used as a loop or dragged into Padshop and used as the basis of a synth. The possibilities are almost endless. I've tried to recreate the atmosphere of each site in the tracks, and let the sounds found on site dictate the feeling. For example, music drops of water inside a burial chamber become percussive loops; crows, cows, wind and damselflies become drones, even overhead aircraft were utilised.

As a project about atmosphere, time and space, *Megaliths* is less obviously musical and rhythmic than, say, *The Journey to the Stones*, but there are similarities in terms of the way the music unfolds in an unhurried fashion, with underlying drones suggesting a particular emphasis on continuity in the landscape. Remarkably, one track from the album, 'Boscawen-un', named for a Bronze Age stone circle near Land's End, has been released in a longer seventy-minute version whose lugubrious, layered drones shift subtly and hypnotically. The impression given is of a bleak, depopulated landscape, open to the elements, with the wind creating echoing eddies of sound that create a deep resonating undertow. It is only on tracks like 'Avebury' and the elegiac 'Pentre Ifan', where we can hear sheep bleating, dogs barking in the distance and the ubiquitous cackling of crows, that we are encouraged to think about the peopling of these sites and how they have been subsumed within a post-productivist rural landscape.

Powell acknowledges his musical engagement with the stones comes at a time when this is becoming de rigueur and could easily be dismissed as a tokenistic nod towards questions of antiquity. He stresses, 'my Instagram feed is full of people singing folk songs on the moors with antlers on their heads, dressed like the cast of the Wicker Man', continuing by voicing his frustration that his once 'isolationist hobby' has suddenly become popular, with some secluded paths becoming rather too well-trodden. Here the pandemic and lockdown obviously played a role, Powell arguing, 'people were so desperate to get out the house and back to nature, to reconnect to the earth'. Powell is also suspicious of those who link the stones to ideas of Englishness or even Britishness. As he puts it, 'the mountains, coastlines, moors and standing stones that inspire me are far older than England, Wales and Europe' continuing:

> I'm fascinated by cosmic horror, the sheer enormity of time and space that makes our blinkered reality so very insignificant. The weight of rain and ice that has sculpted the jagged rocks of Cwm Idwal and Llyn Y Cwn over millennia – in comparison, the modern digital age is a mere blip on the timescale of the earth. We are nothing. We are irrelevant.

Emphasizing notions of deep time, Powell gives short shrift to those suggesting the construction of stone circles constitutes the first evidence of nation-building. Here, some have twisted archaeological evidence to argue the construction of Stonehenge represented a pivotal moment of national co-operation, noting that its stones may have been transported the length and breadth of the country.[18] But Gordon Barclay and Ken Brophy argue that headlines suggesting this cooperation was a 'Neolithic Brexit' at a time when the island was isolated from the continent are way off the mark. As they stress, there was probably no awareness of separation when the stone circle was constructed, and the community leaders involved in its construction would have had no sense they were involved in the making of a nation.[19] Barclay and Brophy also criticized the decision of English heritage to project images of Queen Elizabeth onto Stonehenge to celebrate her seventy years on the throne in May 2022, as if the stone circles are linked to a monarchy whose origins begin in Neolithic times. Such attempts to claim the stones for the nation, for the monarchy or for Brexit are the antithesis of Stone Club's inclusive attitude and Powell's emphasis on deep, geological time. But such revisionism can also be a source of artistic and musical inspiration, with the fanciful equation made between stone circles and nation-building feeding into the weird world of Megalithic Transport Network.

Great British railways and the Megalithic Transport Network

Despite apparent nominative determinism, Martyn Stonehouse is a musician whose interest in stones is relatively recent, coinciding with the recent resurgence in prehistoric perambulation and weird walking. Born in 1986, Stonehouse started experimenting with tape recorders and keyboards in his teens, volunteering at a community arts space at the age of sixteen and ultimately managing a multimedia studio space. Alongside production and education, he released music as DATAstream, a celebration of 1980s and 1990s studio technologies, perversely issuing tracks on obsolete floppy discs at a time when no one could play them. Born in Chesterfield, Stonehouse now works at the local college after having lived in Manchester for a few years, and though always aware of living adjacent to the Peak District, did not really begin to explore it until he developed his recent fascination with stones:

> I didn't really think about stone circles until four or five years ago. My partner mentioned Arbor Low, which I had heard of but never visited, and some other sites down in Cornwall. It got me wondering if there were more nearer home in Derbyshire, and there was a sudden realisation there's loads of these kind of immovable objects in the landscape that are all linked and yet also separate, part of a larger idea. They're beautiful works of art in their own right but what's their purpose? We don't know. And I think that mystery inspired me to think outside the box a little bit further than I had been doing . . . I started to think of the stones as a network, like a network map of ancient Britain.

Slowly, he connected this to a parallel interest in rail travel informed by memories of regular family trips from Derbyshire to relatives in Essex in the 1990s, and his belief that the railway network is now in 'a terrible state', Victorian-era infrastructure patched-up and given a veneer of modernity through regular acts of regional rebranding and new livery.

From this, the idea for Megalithic Transport Network began to solidify, a musical project based on the conceit that the 1970s witnessed the construction of a major new rail branch linking Neolithic sites across the Peak District to a national rail network:

> I analysed the iconography and the fonts and type-based design of British Rail in the late 1980s and I liked what they were doing with regionalisation. They had their regional railways, which were the smaller trains that went out into smaller stations, you know, little branch lines and things like that. So they had these kind

> of like 'hooks' into small communities, part of a larger network. And I guess in my head the Megalithic Transport Network was one of those small networks, joined to the National Rail Network to serve the community's interests, the scenario being that people suddenly had this need to be in the sites of our ancestors.

Working from this admittedly strange premise, Stonehouse has released three six-track EPs documenting the successive phases of development of the Megalithic Transport Network, each accompanied by network maps and brand imagery (Figure 6.6), as well as an EP about the archaeological excavations that preceded the third stage (*Phase Three Site Survey: Excavations on Harthill Moor*). Another EP concerns a branch line extending to the copper mines of Alderley Edge in Cheshire (*Engine Vein*).

These releases hypothesize that if the Beeching Cuts had not happened, and the now-defunct Derbyshire lines that linked Matlock to Buxton and Manchester were kept open, these would have been extended in a period dating from roughly 1976 to 1996. The recording process Stonehouse describes is quite minimal: he uses a small desktop AE modular synthesizer, a few reverb effects units and another KORGMS2000 synthesizer that sits alongside these, which are all plugged straight into a tape recorder, with no overdubbing, before it is digitally transferred. Field recordings are used, mainly as inspiration rather than

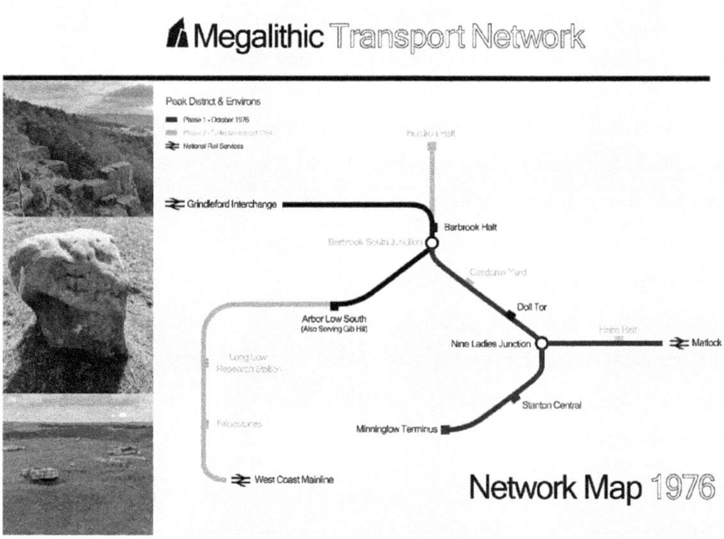

Figure 6.6 Megalithic Transport Network, Network Map 1976. Courtesy: M Stonehouse.

appearing in the recording. Stonehouse suggests field recording encourages him to sit and listen, soaking up the ambience:

> When you're recording something, it encourages you to be quiet and just to think about it. And then you can appreciate where you are and what's going on, which you wouldn't get if you're just walking around with someone having a chat. So there's something nice about the process of recording sound in locations that enables you to understand it a bit better or perhaps appreciate it more.

The output is certainly impressive, each EP having a distinct sonic aesthetic but with a clear progression between them, with the basic formulae – arpeggiated modular pieces that build slowly to evoke both the majesty of the stones and the coming of the railway – developing across the releases as the network expands. By Phase Three, for example, the music is more rhythmic and beat-driven, acknowledging the introduction of new electric rolling stock and the connection of High Peak Junction to Buxton via Nine Ladies Junction. As Stonehouse argues, each release offers an imaginary soundtrack giving a sense of how it might feel to be journeying by train to the ancient sites of the Peak District (Figure 6.7).

This gradual expansion of the Network was (fictionally) deemed to be possible after a site survey in March 1984 of the prehistoric landscape around Harthill

Figure 6.7 Martyn Stonehouse at Doll Tor, Derbyshire. Courtesy: M Stonehouse/Lizzie Henshaw.

Moor, a conceit that allows Stonehouse to explore the relationship between the ancient landscape and the incursion of modern technology:

> I wanted there to be this element of discovery as part of the proposals to build these places. They're not only saying there's a stone circle over there, we need to put train lines to it: they also need investigate what's going on in these places before the trainline goes through it. And *The Excavation of Harthill Moor* was part of the part of that plan, a survey to see whether it's viable to put a new line in, and what they find along the way. And Harthill Moor is a really interesting place. Harthill's got a Roman Fort, Bronze Age barrows and there's the Nine Stones stone circle. There's also Robin Hood stride, which is this amazing collection of gritstone boulders and then you've got the 13th century hermit's house. There's just so much going on.

This development of a coherent backstory grounded in local histories and geographies, the acknowledgement of the interrelation of ancient and modern and the pitch-perfect invocation of the sonic and aesthetic motifs of an earlier decade all mark this out as a quintessentially hauntological project. The question is why Stonehouse has adopted this approach when the elaborate conceit of the Megalithic Transport Network makes his music potentially less accessible (and it is notable that he has issued only limited cassette runs of around thirty copies of each phase, despite the quality of the music suggesting a potentially much wider audience). His view is the project is giving him a new understanding of how ancient sites ought to be treated, something he wishes had happened sooner:

> These places feel reverent. When you stand there and you think about it and the people that have been there before, you know, they feel like they need your attention and you should be paying attention to them. I'm not a religious person but I do think I'm very open-minded.

Perhaps contrary to the premiss of Megalithic Transport Network, Stonehouse is adamant that he is not promoting the construction (or revival) of railways near ancient monuments. Neither, he tells me, is his project any sort of comment on plans to build a major road tunnel under Stonehenge (though some others have interpreted it that way). In fact, at one point he tells me he doesn't want to have to explain why he is doing what he does: the fact is it excites him, and if it doesn't translate to a wider audience that doesn't matter too much to him. He sighs, shrugging his shoulders: 'I'm not trying to say anything in particular, I suppose'. He pauses before elaborating: 'the purpose of the project isn't to push any particular agenda or point of view, it's more about exploring and sharing an enthusiasm and passion for these places, regardless of whether there's a wide listenership or not'.

Given any understanding of the meanings and uses of prehistoric megaliths disappeared long ago, it is perhaps not surprising that hauntologists like Stonehouse are enthusiastically participating in the fabrication of new myths about the stones. Ultimately, ideas that they were built by supernatural beings, served as the focus for folk rituals or are portals to malevolent forces cannot be proven or disproven, so playing with these ideas has appeal for those who want to question ideas of social progress or perhaps point out the shallowness of contemporary culture in its privileging of digital content over the materiality of the numinous landscape. In the final analysis, megaliths, stone circles and dolmens are always in process, their meanings reshaped and supplemented over time, and their ritual role re-enacted or repurposed by each successive generation.[20] Music-making has always been part of that process, and hauntological takes on these historical sites clearly have a role to play in encouraging the widest possible participation in the cultural, collective and therapeutic rituals of visiting prehistoric landscapes.[21] But more than this, they question ownership, suggesting it is dangerous to allow the stones to become the property of a cultural elite: as Stone Club suggests, heritage is there to be enjoyed by the many, not the few.

Further Reading

Burl, Aubrey (2000). *The Stone Circles of Britain, Ireland, and Brittany*. Yale University Press.

Burnham, Andy (2018). *The Old Stones: a field guide to the megalithic sites of Britain and Ireland*. Watkins.

Doyle White, Ethan (2014). 'Devil's stones and midnight rites: Megaliths, folklore, and contemporary pagan witchcraft'. *Folklore*, 125(1), 60–79.

Edensor, Tim, & Brophy, Kenneth (2023). 'The potent urban prehistory of an ancient megalith: The Kempock Stone, Gourock, Scotland'. *International Journal of Heritage Studies*, 29(1–2), 81–96.

Nolan, Claire (2019). 'Prehistoric landscapes as a source of ontological security for the present day'. *Heritage & Society*, 12(1), 1–25.

Pitcher, Ben (2016). 'Belonging to a different landscape: Repurposing nationalist affects'. *Sociological Research Online*, 21(1), 77–89.

Pitcher, Ben (2022). *Back to the Stone Age: race and prehistory in contemporary culture*. McGill-Queen's Press.

7

Strange Ruralities and Folk Geographies

If hauntology is defined as the presence of absence and a concern with the eerie aspects of landscape, for Fisher, the weird is, to the contrary, constituted by 'the presence of that which does not belong'.[1] For him, it signals something that has no right to be: 'a weird entity or object is so strange that it makes us feel that it should not exist, or at least it should not exist here'.[2] For some, the eerie and the strange are part of the same problematic: if we live in haunted times, we must also live in strange times.[3] For example, in his analysis of recent landscape writing, Roger Luckhurst suggests the contemporary turn to the 'weird' can only be understood in relation to Brexit, which he theorizes as a turn away from the global and a putative return to a local imagined as stable and solid. For him, the embrace of the weird is Brexitland's 'dark shadow', a response to the vote by those who are 'seeking revenge' on those who have constrained meanings of England and Englishness within narrow definitions of rurality and locality. For Luckhurst, place is always convoluted, muddy with myths that generate alternative understandings of what it means to be English. Attempts to identify, and celebrate, weirdness in the English countryside are therefore read by Luckhurst as a progressive attempt to subvert national myths, diving into a past that cannot readily be routinized by the heritage industry.[4]

In film too, it is clear that the prominence of recent British horror films that obliquely explore questions of national identity and heritage via a focus on the sinister dimensions of landscape – including *Enys Men*, *In The Ground*, and *Starve Acre* – is connected to what J. M. Tyree terms 'the return of the repressed' in political consciousness.[5] Here, films that depict the English countryside as inherently weird, dwelling on the persistence of occult and pagan practices in the present, play into ideas of the rural as 'out of time', with the Gothic grotesque an important trope challenging the mythos of the green and pleasant land that was often to the fore of pro-Brexit imagery.

Hauntologists and psychogeographers have likewise long been drawn to rural sites – hill forts, stone circles, cairns – where the sonic reverberations of pagan pasts and occult ritual collide with the sounds of illegal '90s rave parties, nefarious nocturnal couplings and drug-induced violence. In such places, the countryside appears oddly out of joint with ideas of the ordered English countryside, the pastoral overlaying a barely repressed landscape of criminality, depravity and horror. Considering this representation of the rural as 'anti-idyll', geographer Tim Hall notes many instances where the picturesque tranquillity of the countryside is accordingly played down in favour of an emphasis on rural life as bleak, grim and isolated.[6] Such myths are often consolidated by the depiction of rural 'folk' as unmodern, suspicious and capable of enacting extreme violence to conserve rural ways of life, with folk horror a genre accentuating the distinction between the cultured metropolis and the superstitious countryside.[7]

In his definitive history of the genre, Adam Scovell defines folk horror as involving a horrific event or summoning occurring in a landscape that appears isolated and backward. While Scovell stresses the events that unfold often involve the presence of the supernatural or the occult, sometimes the dangers and terrors portrayed are more earthly, with the rural a site where those who suffered childhood trauma or abuse seek revenge on their perpetrators (e.g. see Matt Holness' 2018 queasy psychological horror film *Possum* or the 2013 Warp TV series *Southcliffe,* in which a shooter runs amok in a small town on the North Kent Marshes). Either way, folk horror infuses the landscape with an atmosphere of creeping dread and a suspicion that all might not be as it first appears. Just below the surface, personal, social and spiritual anxieties are buried, waiting to be exhumed by the unwary.

This depiction of the weirdness that lurks in the countryside was arguably first articulated in the 1970s horror films that were a reaction to the hippy idealism of the 'back to the land' movement that emerged at a time of oil shortage, hyperinflation and power cuts. But in the context of the post-Brexit political crisis, folk horror has become something of a twenty-first-century cultural phenomenon: witness the acclaim heaped upon, for example, Mark Jenkin's (2022) mind-bending *Enys Men,* which has a stunning soundtrack combining 1970s radiophonica, tape loops and ambient noise, performed live at a number of screenings by the Cornish Film Unit (Jenkin with Dion Star). J. M. Tyree suggests this film, set in 1973 but moving between past and present events, resists any idea that England is an island nation that must repel invasion by suggesting that the English themselves *are* the invaders, continually disturbing fragile forms of life. In *Enys Men*, scenes that show lichen growing on a woman's body depict an 'ecological uncanny', a form of body horror that suggests the divides of nature and culture are not as clearcut as the pastoral curatorship of the land might imply.

The resurgence of interest in the tropes of rural horror has witnessed many musicians revisiting the folk horror canon of the 1970s, notably the 'holy trinity' (*Blood on Satan's Claw*, *Witchfinder General* and *The Wickerman*), but also numerous 'wyrd' TV series, most notably *The Owl Service, Penda's Fen,* and *Raven.* As Scovell points out, the fact many of these series, not repeated on TV since the 1970s, are now available for viewing on YouTube is not coincidental in their citation by musicians who have, on occasions, based entire albums around some of these formerly obscure programmes. Yet this hauntological enthusiasm for self-consciously citing and remixing the folk horror of the 1970s has been evident across a much wider range albums fusing electronica with acid folk and post-pastoral psychedelia, including releases from Cobalt Chapel, The Hare and The Moon, The Night Monitor, The Rowan Amber Mill and The Northern Lighthouse Board to name but a few.

Stephen Price's *Year in the Country* project has provided a formidable resource for those exploring the geographies of folk horror. Since 2014, his website and a series of book-length compendia, have provided a guide to the TV, film and literature that has contributed to the uncanny sensation that the English countryside is not all that it first appears.[8] This has included discussion of many classic films in the folk horror tradition but also more obscure writers and performers operating in the space between experimental counterculture and the popular mainstream. From Boards of Canada to *Bagpuss*, *The Detectorists* to *The Day of the Triffids*, Prince has been culturally omnivorous in his attempt to find the roots of the rural wyrd by making connections between psychedelic pastoralism, folk culture and hauntology. His own forays into fiction – the novellas *The Corn Mother* (2020) and *The Shildham Hall Tapes* (2021) – also play with the traditions of the folk horror genre, in both cases purporting to be novels based on the script of abandoned films from the 1970s. Significantly, both novellas accompanied albums which were the imagined soundtrack to these films: for example, *The Shildam Hall Tapes: The Falling Reverse* (2018), featuring Sproatly Smith, Field Lines Cartographer, Vic Mars, Pulselovers and others, 're-assembles' fragments of audio from the original, albeit fictional, film.

Alongside these books, imagined soundtracks and blog posts, Prince curated a formidable series of compilations between 2014 and 2020 with contributions from many of these acts, alongside others prominent in the wider hauntological scene including Howlround, Keith Seatman, and Listening Center. One of the regular contributors was The Heartwood Institute, a musical project by the prolific library music composer Jonathan Sharp, responsible for nearly twenty albums since 2015, many tapping into the themes of folk horror. A largely self-taught guitarist, Sharp played in industrial-goth bands (notably Hexedene and Novus) before starting to compose content for virtual instruments and making library music in the early 2000s. He started The Heartwood Institute as a side-project in 2015, fusing dark ambient, radiophonica and neo-folk to evoke the unease that sometimes permeates rural landscapes. Sharp tells me he was particularly inspired by the unsettling urban soundscapes of Burial, wanting to produce a rural equivalent. He also claims affinity with Pefkin and Bell Lungs, performers pushing folk music in new directions through their use of synths and samplers. The Heartwood Institute is hence a musical experiment evoking the weirdness of the rural:

> I love Bob Fischer's description [in *Electronic Sound*] of what I do as analogue synths meets agrarian darkness. That's 100% what I'm aiming for. Fieldwork and location recording plays a very large part of the overall sounds I use. I've captured Morris dancers and turned them into a sampled percussion/drum set. The lychgate of my local church has been warped into several different sounds. I'm just trying to create my own sound world from my own environment and my equipment.

These found sounds and influences emphasize the sinister elements of landscape. 2023's *Mist over Pendle*, for example, is an imagined soundtrack to the 1976 BBC TV film dramatizing the 1612 witch trails held near Colne in Lancashire, the cover depicting Cathryn Harrison as Alizon Device, one of ten killed by overzealous witch-finders. Slurry radiophonica, movie samples, and ominous synth chords all combine to create a faux 1970s soundtrack which is distinctively hauntological in its citation of the past, its reproduction of old sonic technologies and the creation of an atmosphere of suspense. *The Witchcraft Murders* (2021) ploughs a similar furrow by offering an 'audio investigation' of the Lower Quinton and Hagley Wood murders in Warwickshire, which involved the grisly 1945 discovery of a male body on Meon Hill with a pitchfork through its neck and a cross inscribed on its chest, and the similarly macabre murder of 'Bella' in 1943, her pregnant body found stuffed inside a tree trunk by children playing in the Clent Hills. A 2020 release, *Witchseason*, offers three terror-infused tracks inspired by ghost stories, while *Secret Rites* (2018) samples liberally from the faux mondo documentary *Witchcraft 70*, a film fusing the sexploitation genre to stories of black magic.

Just what would provoke a musician to create concept albums exhuming the disquieting traces of the rural past? For Jonathan Sharp (Figure 7.1), it is not just the memory of the folk horror and TV of the 1970s, but the experience of growing up in Cumbria, a county steeped in tales of the arcane and the occult. For example, he recalls being told as a child that a bizarrely painted house in a nearby village was 'the witch's house': as he tells me, 'you don't have to dig very far into local folklore before you run into stories of pacts with the devil, hands of glory and witchcraft in general . . . It's a very rich vein to mine'. Recounting such stories, Sharp argues the Cumbria he grew up in was somehow closer to the past than other, more modernized, places, boasting a rich sense of place and time. He notes that as a child he spent a lot of time with his grandmother, born in 1900 and 'very much connected to a different, older way of life that was far more rural', and 'spoke in a very broad Cumbrian dialect too'. His 2022 album *Land of The Lakes* is dedicated to her, and is a relatively bucolic album, in thrall to the soaring Lakeland landscape. The album, inspired by finding a pre-First World War picture book on the Lake District, depicts Cumbria as the quintessence of

Figure 7.1 Jonathan Sharp, Cumbria, 1975. Courtesy: the artist.

upland Englishness, rugged but charming, bound together by timeless custom and ritual. But Sharp acknowledges this is fast disappearing:

> In terms of landscape, it makes me sad to see how people now just treat it with no consideration: the Lake District isn't just some theme park for your entertainment and to be treated with no respect, it's a living working environment that suffers from people treating it like a festival site where they can abandon anything they can't be bothered to take home. And while I totally support the right to roam, I probably approach it from a slightly different viewpoint. I grew up and live in a pretty rural area surrounded by farms. So I feel you need to be aware of treating the land and the animals on it with due respect.

It is tempting to wonder if there's also a link between Sharp's frequent invocation of folk horror and the rural gentrification that has left one in ten of all houses in the county empty for some or most of the year, second homes for the metropolitan elite. The idea of outsiders failing to understand 'rural ways' is certainly a frequent trope in folk horror, a veiled comment on the threats of inward investment, touristification and middle-class invasion. Sharp himself admits to being 'politically motivated from a protest point of view', expressing an interest in counterculture and the history of protest. He puts much of this down to living through the Miners' Strike of the early 1980s, recalling 'I don't think I'd seen such widespread or violent protest up to that point. I have an abiding memory of travelling down the M6 and seeing Police pull over cars and vans full of strikers . . . That certainly effects how you view the right to protest'.

The link between music and rural protest is long established in the folk idiom, but the connections between folk horror, hauntology and class tensions in the countryside are perhaps less noted. Yet as James Thurgill writes, the true horror of the countryside emanates 'not from the social inadequacies or archaic customs of rural populations but . . . a lack of investment in, and an estrangement from, central governance and metropolitan hubs'.[9] Here the invocation of a folk community living in opposition to a modern, gentrifying middle-class plays up stereotypes of town and country, drawing attention to the processes supporting the uneven production of space. Though rarely discussed in these terms, there are important links here between gothic Marxism and the idea of rural spaces – and bodies – being transformed into commodities.[10] As Thurgill argues, rural horror actually exists 'in shrinking rural economies; the closure of primary resources such as hospitals, doctors' surgeries, libraries and schools; restricted communications, diminished public transport services as well as internet and cellular 'blackspots'; the brutality of livestock farming and its subsequent body count'. Even in the wealthy, gentrified countryside, there are evidently still many living on their uppers.

Julian Holloway is someone who has thought more than most about these issues given his academic work on the cultural geographies of the supernatural at Manchester Metropolitan University is pursued alongside his music-making as a member of Flange Circus, a self-styled dark wave, neokraut and folk horror trio (Figure 7.2). Holloway confesses to being interested in what he calls survivals – 'those bits of the past that persist despite the apparently rational and planned spaces we live and work in . . . things that defy rationality and we're never sure where they originated from, what they might mean, or how they survive'. This is manifest in Flange Circus' evocations of the rural wyrd: across three albums, and several singles, they have explored the grotesque and menacing dimensions of the rural through dark slabs of drone, with *The Rural Eerie* (2021) being a high point, a remarkable summation of the anti-idyll through music, commissioned poetry and an accompanying film. Tracks reference the events of the past that still haunt many villages: one track – 'Desolation' – is about 'lovely, lone and still' village of Eyam which, in the summer of 1666, experienced a plague outbreak killing seventy-three inhabitants, the track featuring stanzas from Mary Howitt's (1827) poem 'The Desolation of Eyam'. On the same release, Mark Pajak's grim tale of removing dead chickens from an industrial-scale broiler house as a

Figure 7.2 Flange Circus. Courtesy: Julian Holloway.

sixteen-year-old suggests continuities between the terrors of the past and the labour of the present. Yet beyond these allusions to gothic Marxism, there is also emphasis on the idea that rural settlements are particularly susceptible to the elements: Andrew Michael Hurley – author of *Starve Acre* (2019) and *Barrowbeck* (2024) – recounts the 'purgative coldness' that grips a valley winterlong, only grudgingly released in the springtime. Helen Darby's gothic-tinged poem likewise speaks of the frozen ground of Derbyshire and the memories of the last Harvest that linger through the dark winter, while Sarah Hymas talks of the driving rain making everything slippery and dangerous: her poem is backed by the sound of an accordion rising and falling as if it were the wind, keyboard drones and swirling ethereal wails gradually growing in intensity to produce a sense of tension only released as the track ends.

The Rural Eerie is an absorbing sonic experiment, and one that amply illustrates the power of what Holloway terms *resonance*: a performative attempt to recreate the geographies and histories of place that eschews mimesis to focus instead on affect, atmosphere and mood. Here, the listener is guided by liner notes, elaborate backstories and imagery that summons ideas of the sinister rural. Irrespective, the music itself conveys the 'spatial vectors' of the rural wyrd. Half-heard and half-known, it creates a soundscape both familiar and strange because we can't be sure what we are listening to. As Holloway describes, 'through their disquieting and eerie atmospheres, these sonics allow us to create and practice ruralities that may have not existed or may never exist in some alternative past or future'.[11] His summation is that 'folk horror says a lot about landscape and the eruption of something unknown that can both delight and scare us' suggesting that because 'our lives are given order by routines, authorities and forces mostly beyond our control, if we revel in the unknown and the enchanting we might be able to see the world a little differently . . . and progressive change might be summoned'.[12]

Downland dreaming

Folk Horror appears particularly potent when describing the isolation of villages and settlements in upland Britain and the Celtic fringes, its wild moorland and backward villages constituting 'the periphery of the periphery'. Many recent books, films and musical evocations of wyrd rural happenings accordingly allude to the 'sparsely occupied areas' where industries have died, invoking the retreat of capital and the persistence of the North/South divide. Robert Edgar speaks of these as spaces where 'a form of civilisation has risen and then, as economic fortunes have changed, has been left to rot', making connections between folk horror, the Brexit leave vote and the emergence of 'skewed belief systems' resulting in abandonment rather than isolation per se.[13] The representation of these spaces as uncared for and uncaring resonates with metropolitan audiences, whose own indifference to these spaces has, in part, caused their decline.

The established tropes of folk horror, which offer a skewed critique of contemporary capitalism, are then most convincing when referring to landscapes far from the Southern pastoral landscapes taken to epitomize national values. This colours the expectations of audiences, of course. Flange Circus, for example, have used field recordings from a variety of locations, North and South, in their rural eerie, but according to one reviewer, their work 'hangs heavy with the air of the Northern climes of this sceptred isle, with the spoken word conjuring up the past and present of the North-West – of the Lancashire landscape carved at and reformed by the long-faltered advance of the industrial revolution'.[14] This suggests that certain well-known place myths – of witch trials and plague pits, public executions and deserted medieval villages – predispose us to associate folk terror with the landscape of the North. In contrast, the Home Counties seem to fit less well to the classic tropes of folk horror, offering prettiness and prosperity rather than creepiness and capitalist abandonment.

Yet there are exceptions, with some of the settled and bucolic landscapes of the South also harbouring memories of horrific happenings and dark deeds. Justin P. Hopper's (2017) *The Old Weird Albion* provides an exemplary scrutiny of these, an account of walking the South Downs from Winchester to Beachy Head which eschews the usual conventions of travel writing to explore the entanglement of autobiography and landscape. Throughout, Hopper fixates on the mysterious 'lost places' and shadowlands that remind him of his own mortality, including standing stones, hill forts and pagan sites. While he was born in the US, Hopper's grandparents lived in Steyning, near Brighton: his grandfather's first wife Doris leapt off Beachy Head, Eastbourne, surviving for another year before dying of her injuries. This tragic backstory becomes the context for Hopper's poetic psychogeography of the South Downs Way. One memorable chapter focuses on the ancient hillfort on top of Chanctonbury Hill, West Sussex, which Hopper describes as 'one of the thinnest places', a place of hauntings and levitations where 'dead voices gather' and memories linger 'like snagged wool' (Figure 7.3). Hopper argues this landscape remembers 'every

Figure 7.3 Chanctonbury Hill, West Sussex. Photo: author.

footfall and wheel-roll, and bears each of these as lines on its face', tracing a history of the hill that encompasses druidic ceremony, the occult, the murder of a London businessmen and UFO sightings.[15]

Folk horror hence is an important motif in much of Hopper's writing, with the pagan past threatening to return via the collapse of past and present time.[16] Local myths say if you run counterclockwise around Chanctonbury Hill seven times, it will summon the devil. While Hopper does not test this out, on May Day on the hill, he sees his long-dead maternal grandmother manifest before him and falls into a fever dream, his reverie broken only by the ringing of Morris dancers' bells as a local group enters, dancing the 'Bonny Breast Knot', a ritual designed to 'mark time' and welcome spring.[17] A later, embellished, reading of this chapter, accompanied by psych-folk performer Alison Kraus and Belbury Poly, was issued by Ghost Box Records in 2019: this further accentuates the ethereal qualities of the landscape, Kraus' wordless vocals and chanting on tracks like 'Breath' readily summoning visions of witchcraft.

In general terms, despite its proximity to London and its relative wealth, Sussex has been considered the antithesis of the modern metropolis, a place where rural folk went about their business with 'little hurry, noise, smoke or stench'.[18] This reputation has attracted writers and artists in large numbers, drawn by the timelessness of the Downland landscape of chalk, wood and water. Historian Peter Brandon suggests that much of the creative activity in Sussex – and indeed England – in the inter-war years occurred within ten miles of Firle Beacon, from where one can look over the rolling Wealden landscape but also have a view of the English channel. Rudyard Kipling, William Nicolson and Frank Newbold all lived near here, while members of the Bloomsbury Set often decamped to Charleston, the home of artists Vanessa Bell and Duncan Grant. Victor Neuman established the Vine Press in Steyning. Graham Greene and Patrick Hamilton lived in Brighton, Eric Ravilious taught in Eastbourne. Lewes was a particularly popular town for the literary set, once described by poet Elizabeth Hitchener as the 'most pleasing town in Albion's isle': compact and intimate, set in an amphitheatre of green hills, it was for a short time home to Virginia and Leonard Woolf.

Shirley Collins, who, with her sister Dolly, did much to bring about the 1960s English folk revival, has also lived in Lewes since 2014. Born in Hastings, she met Alan Lomax, the American folk archivist, in 1954, and spent much of the next few years assisting him in collecting folk recordings in the American South. Returning to the UK, she recorded a series of influential albums alongside her sister and, later, her second husband Ashley Hutchings, as part of the Albion Country Band. Retiring from music after her traumatic divorce from Hutchings in 1978, she nonetheless was drawn back to recording in the twenty-first century when sought out by the 'apocalyptic folk band' Current 93, founded by ex-Psychic TV member David Tibet, who also released a collection of her songs on his Dutro record label. Three solo albums have followed – *Lodestar* (2016), *Heart's Ease* (2020) and *Archangel Hill* (2023) – the latter heralded as her last-ever album and received warmly by a younger generation of folk enthusiasts. This was named after the rounded summit of Mount Caburn, to the east of Lewes, which Collins' stepfather (a stable hand) began to call Archangel Hill as his memory began to fail. On the album's sleeve notes, Collins stresses her connection to the local landscape, writing 'whenever I walk Mount Caburn, I give a silent greeting in memory of my stepfather Bill and his horses . . . I've picked sloes there in autumn, sat watching sheep and the occasional chalk hill blue butterfly in summer, but Bill had ridden over it many times in the 1920s, walking horses from Bishopstone to the Lewes races'. Elsewhere, she suggests her music is born out of respect for those who walked the Sussex hills before her:

> Singing English folk songs is as crucial to me as walking the Sussex landscape, where the footprints of our ancestors are everywhere. When I sing, I feel past

> generations standing behind me – and I hope I'm a conduit for them – those farm labourers and their wives who kept the songs going for us. The songs are social history and their beauty and power undeniable.[19]

Her delivery of traditional English folk tunes rolls softly, 'like so many of the Sussex tunes do', she argues. 'They just represent the Downs to me – they rise and fall and they're majestic and they're beautiful'.[20]

Collins' recordings seem a world away from folk horror but unquestionably tap into traditions of folklore celebrating the lives and labour of the rural folk who have worked the land, often in deference to an elite class. Figures from the past drift through her work like ghosts, her summoning of the pre-industrial era added weight by her own memories of rural change. At nearly ninety years old, she recalls living in the South Downs when there was no light pollution to block the view of the Milky Way, a time when life seemed slower, neighbourly and peaceful. Her songs appear timeless, her voice heard as if from an earlier time. This made her the obvious choice as a guest reader for Justin P. Hopper's (2014) spoken word piece *Ley Line*, a recollection of a psychogeographical transect in Pittsburgh which transposes this with the UK and reflects on the way folk songs from Sussex travelled to Appalachia. In the background, sparse field recordings and sound effects add texture, with the lilting pastoralism of the opening track 'Falling Banks' performed by Belbury Poly.

The inclusion of Ghost Box Records' Belbury Poly on Justin P. Hopper's excursion into Sussex is no mere coincidence, with Jim Jupp having lived on the fringes of the Downs in Sussex for some years. In an interview on Hopper's *Uncanny Landscape* podcast, Jupp explains the influence of Sussex on his own hauntological worldview:

> I derive great inspiration from the landscape from the Downs. Hundreds of musicians have been constantly walking the Downs and deriving the same sort of inspiration, but I really do. I do have musical ideas when I'm walking and there's the act of walking which stimulates ideas. I grab my phone and sing a little tune into the phone quite often or just make some notes. The idea of the landscape, and the stories about the landscape and all the stuff we overlay when we look at a vista, that I find interesting. I suppose that's what people call psychogeography, isn't it? I've never really got a handle on what that is, but I think it's the stories we overlay on the personality of the landscape. I like that.[21]

On the 2023 album *The Path*, Belbury Poly take this idea of engaging with the Downland landscape seriously, creating a transatlantic band (Figure 7.4.)

Figure 7.4 Belbury Poly, 2023. Photo: Andy Sapi, courtesy: Ghost Box.

featuring Justin P. Hopper as narrator and Jesse Chandler (of Midlake) on flute and keyboards to tell the tale of an American tourist who follows one of the 'old ways' out of town, finding the modern landscape of motorways and trunk roads leads him ultimately to moss-laded paths tangled with memories and ghosts of the past, flickering faeries glimpsed in the fading light.

The album dispenses with the samples and Radiophonic flourishes of earlier Belbury Poly releases for a more organic sound fusing 1970s prog rock and pastoral psych, with occasional sitars invoking memories of bands playing free festivals on the Downs.[22] Musically it offers a distinctive reinterpretation of folk music, one not faithful to the musical traditions of Sussex but approaching the Downs through the psychedelic imaginations of the late 1960s and early 1970s. Jupp explains:

> The musical ideas draw on folk music and pastoral music and even early music, but . . . I didn't grow up with those things. I'm not steeped in them. They're as received through the TV when I was a kid . . . I heard those things and they were mysteries to me because I didn't know what they were. Whether it was on film soundtracks or weird kid's programs, there's a mystery that's not there in the authentic thing. As much as I enjoy traditional folk music I prefer, acid folk, eerie folk from the 60s . . . That's the received stuff that interests me, because that's where there's a disconnect: it can lead you somewhere else.[23]

As Jupp recalls, the album was based on the idea of 'an American, maybe an academic, wandering around, obsessed with British landscape and getting overwhelmed by the weirdness of it all'. On tracks including 'Highways and Byways' and 'The Wrong Spot' we hear Hopper becoming more disoriented, lost in a hallucinogenic world beyond his comprehension, the path having taken him away from who he was – or thought he was.

Jim Jupp suggests much is left vague on *The Path* so the listener can make of it what they will, but there is little question that it invokes the rural uncanny: the sense 'you can be in the countryside, in a gorgeous place, and you suddenly get this odd mood of disconnected isolation'.[24] In this regard, *The Path* was an indirect sequel to the preceding Belbury Poly album, 2020's *The Gone Away*. This tapped into the faerie mythology of Sussex, and the idea the Downland landscape shelters 'little people' who, according to some, were the last survivors of a Neolithic race. Part of the richly layered folk narratives associated with the South Downs, historian Peter Brandon argues that the superstitious villagers of Sussex believed in faeries, dryads, nymphs and fauns late into the nineteenth century. By the twentieth century, however, these myths were held to less out of belief, and more in the hope that talk of faeries and satyrs would help keep alive a version of rural England that was rich in folklore. The poet Charles Dalmon, friend of Noël Coward and later member of Moseley's Union of Fascists, devoted much of his 1900 *Flower and Leaf* collection to stories of faerie harpists playing dancing tunes behind the 'secret doors in the woods', while Kipling's (1906) *Puck of Pook Hill* finds two children summoning a faerie by performing a scene from *A Midsummer's Night Dream* three times on a hot summer's night, the story set near Burwash in Sussex. The composer John Ireland also reportedly found himself surrounded by singing faerie children on a visit to Chanctonbury Hill in the 1920s, writing to his friend, the decadent horror writer Arthur Machen, about the experience. Machen replied on a postcard, 'So you've seen them too?'

Following the Great War, faerie tales were kept alive by those who had lost sons and daughters in the war. Representations of fairies transitioned from the demonic or elemental towards the delicate, winged childlike figure close to the Disney version we know today. But while the faerie myths of Sussex persisted into the inter-war, they could not survive beyond the Second World War, with some logically suggesting the faeries departed because of the invasion of the motor car, the sprawl of suburbia, light pollution and the incessant noise of mechanized farming. As Jim Jupp describes, there's a common trope in faerie law that somebody heard a noise in the middle of the night, and 'saw all the faeries walking off towards the coast where they sailed off', leading to the album being called *The Gone Away*.

A concept album about faeries disappearing from the Sussex Downs is both far-fetched and fantastical, but taps into hauntological themes concerning the way that some histories have come to be forgotten or rewritten. Indeed, Jupp argues there are good reasons to revisit the cultural symbolism and meaning of the faerie within English folklore at this moment, with the present absence of faeries on the Sussex Downs speaking to a range of contemporary anxieties:

> The whole topic seems very current . . . I don't really know why. I think it certainly chimes with environmental concerns and issues, ideas that it may be time for nature to take revenge or take things back. Some of the nutty theories about coronavirus were that the lockdown made the atmosphere better, wildlife comes back, and it's almost like, oh, hang on, is somebody arranging for us not to do all this bad stuff anymore? The idea of the fairy, especially in folklore, where they're quite malevolent, is they'll protect a field, or a wood, or an animal . . . and woe betide the person that crosses them, or does the wrong thing![25]

The Gone Away exists simultaneously in the Sussex of the 1890s and the present day, its otherworldliness reflected in music recalling both ancient folklore and modern environmental anxieties through a unique fusion of Tolkien-stained prog, modern electronica and quaint neomedieval pastoralism (one track, 'Sticks and Stones', is perhaps best described as medieval dubstep). Rather than any slavish retelling of fairy myths, this is an allusive use of folklore, a creative deployment of the non-rational, hidden and magical that infuses the landscape.[26] Even if one does not believe in fairies, there is clearly something powerful and important about the stories we make of them.

The ritual year

The fact many musicians and performers working with hauntology are interested in folklore speaks to a concern with reversing the disenchantment associated with capitalism, where commodity fetishism has taken over not just the economy but society as a whole. The search for the magical – or what Jane Bennett refers to as 'the extraordinary that lives among the everyday and familiar' – has subsequently been thwarted every time practitioners have proposed ways to enchant the banality of ordinary life, whether via surrealism, mysticism, psychedelics or eroticism.[27] It is here that hauntology potentially offers new tools for re-enchantment, not just by highlighting the weird and wonderful, but by rupturing our normal 'sensory-psychic-intellectual dispositions'.[28] Interest in, and enthusiasm for, folklore is one part of this, but so too is a rejection of industrial clock time in favour of an awareness of seasonal customs and cycles of nature. Renewed contemporary interest in pagan seasonal festivals, oriented around the four key solar events (solstices and equinoxes) is evidence of this. So too is observance of the pivotal festival days punctuating the traditional agricultural year: Candlemas, Lady Day, May Day, Midsummer, Lammas, Michaelmas, Halloween and Christmas.

Reflecting this, throughout 2023, on the twenty-first of each month, Bonfire Hill – a collaboration between Rebecca Denniff (aka Subphotic) and David Owen (aka Band of Cloud) – released a track. Each was a full twelve minutes long, the resulting 144 minutes of music being the *de facto* soundtrack to *A Year on Bonfire Hill*. This 'aural almanac' was described as a guide to the changing seasons of a place 'not shown on any map' but existing 'in every corner of the country', somewhere fictional but also naggingly familiar:

> Bonfire Hill is somewhere just North of here. It holds a commanding view of the surrounding countryside – the moors, valleys, fields, rivers, streams, woods and, in the distance, the sea. Bonfire Hill can be reached by following the old road. The one your ancestors took. The route they first trod, through the trees, across the plain. There are houses on Bonfire Hill. Some are warm and welcoming, some are quiet and secluded, some are ancient – built into the hillside – some are modern and forward-facing . . . Yet the modern world sits comfortably amongst the ruins on Bonfire Hill . . . The electricity crackles alongside the sparks and flames of the fires. The music oscillates and thrums and the voices sing out and mix with the recordings of singers gone before.[29]

The landscape evoked here shows continuity of occupation, the ancient sitting cheek-by-jowl with the modern, but we are told there is no conflict, for the two are bound together by traditions of music-making. We are hence invited to listen to the folk songs that mark the seasons on Bonfire Hill, from the dark, icy days of winter to the golden sunsets of summer.

Each of the twelve tracks combines field recordings, original composition, and traditional song, underpinned by a swirling mix of folk, ambient and electronic drones. 'January' begins with a rhythmic almost-industrial beat, before we hear the noise of a radio being tuned, giving way to piano notes and a gentle synth arpeggio over which Rebecca begins to chant 'On Bonfire Hill'. She continues to sing: 'January close the door / For the year is turning, turning', before a church organ fades in. We then hear a modern version of an old wassailing song – 'Oh apple tree, bring forth your rosy apples to me' – typically sung on the twelfth night after Christmas to bless the orchards. As the track proceeds, varied samples bleed in, including a disconcerting snatch of Pilot's 1975 soft rock number one hit 'January'. The track finishes with Rebecca singing about a murmuration of starlings in the late January dusk, before the track gives way to the sound of birdsong. The whole is a remarkable and dense sound collage, folk songs threaded through an ambient electronic soundscape that's both ancient and modern. This is the result of careful planning and spontaneous musical improvization, as Rebecca explains:

> We talk about what's going on for ages, that gets us in the mood for whatever the sound is going to be. We've played with keys, and which key we start in, depending on the month of the year or whether it's on the decline or the ascendant. So the ones that are around the equinoxes, there was a key change in them from minor to major, then from major to minor. But when we record, I sit down and will just start creating drones and record them straight away. I'm not rehearsing, we just build up the layers. Then I give all of those textures and soundscapes that I've created to David, who's got all his samples. So he'll listen a few times to the kind of bedrock of what we've built, and he puts the samples in and then it all comes back to me and I sing in response to the samples.

Throughout the twelve instalments, Bonfire Hill allude to the annual cycle of traditional, and sometimes forgotten, English festivals, many with pagan origins. Beltane is marked in 'May', St Brigid's Day in 'February', and Rushbearing festivals namechecked in 'August', the latter referring to the annual laying of rushes and grass on the bare floor of churches in readiness for the winter ahead. In 'September', Rebecca sings of the Hunter's Moon and the 'ripe and bearded' barley 'smiling on the scythe', with John Barleycorn becoming the personification of the process of reaping and threshing, a figure bloodied and killed before the restorative effects of alcohol result in his resurrection. In 'May', the Queen of May steps out with Jack-in-the-Green, part of a pre-Christian fertility ritual in which 'young maidens' stooped to wash their face in the morning dew, anticipating the fecundity of summer.

The artwork for each month accordingly highlights some of the iconography of ancient symbols of the unfolding of the seasons, from an antler-bearing wassailer through to the corn dollies of autumn (Figure 7.5). Church bells, the sounds of Morris dancers' sticks clashing, and samples of hymns drift through

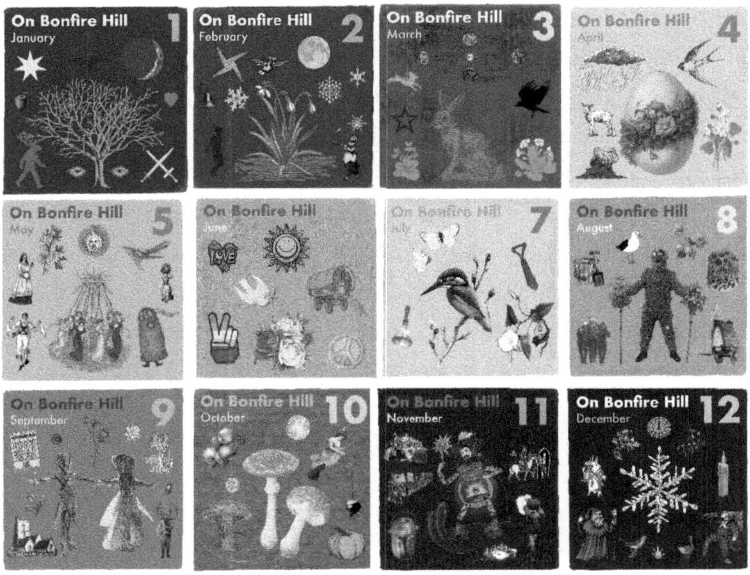

Figure 7.5 A Year on Bonfire Hill. Courtesy of David Owen/Rebecca Denniff.

the music, stressing a connection to time-honoured folk rituals. Songs likewise bear witness to the arrival of the cuckoo, spring lambing, the first swift of summer, the march hare and the kingfisher swooping on the river. A rhythmic geography unfolds in which the natural calendar sets the pace of life, the music mirroring the unfolding of the seasons.

Yet there are hints that nature is not always bounteous, and there are signs of rural unrest here too. The May Queen, for instance, is followed by others not as fortunate as her: 'See the girl with the ribbon in black? / She was stabbed in the back'. At other times, scythes are wielded in anger and miscreants hounded out of town. Here, Bonfire Hill taps into a tradition of folk music that documents betrayal, lost love, adultery and murder. A sinister and unsettling rurality is hinted at, an alternative to the bucolicism often present in mainstream renderings of the English countryside. Here, it's revealing that Rebecca Denniff and David Owen's previous collaboration was as Storm Chorus, who released 'Died for Love' in 2019, an album of murder ballads that traced the tradition 'back to the heart of the lonesome woods'.[30]

Yet the emphasis on the folkloristic in the music of Bonfire Hill, and the repeated use of traditional musical motifs, sits alongside more modern musical references. As well as samples of 1970s music, on occasion the beatless, ambient wash that provides the backdrop to the folkloric lyrics gives way to something of more recent provenance. In 'June', for example, the month when every 'lad takes his lass' to the green grass atop Bonfire Hill, it appears we are in a 1990s rural rave: the lyrics tell us the sun is shining again, and what appears to be a bassline inspired by 'Higher than the Sun' by Primal Scream breaks through. The sample that follows is taken from news coverage of the Castlemorton Common free festival of 1992, capturing some of the moral panic provoked by thousands of ravers descending on the Malvern Hills, including the voice of a local resident complaining about the noise and mess caused by participants. The track builds to a crescendo as an organ plays descending notes on top of the bassline, with a different, female voice proclaiming the rave to be both 'Babylon and Bedlam' before the track ends with a brief sample of Jarvis Cocker, from 1995's 'Sorted for Es and Wizz' (by Pulp), asking plaintively: 'What if we never come down?'

Here, and elsewhere, there is emphasis on rural revolt and radicalism. In 'May', for example, we hear of the importance of May Day in worker's protests: alongside the chirping of birds in the springtime, and talk of ribbons tangled around the maypole, there's a prominent sample of a crowd chanting 'If the workers are united, they will never be divided'. 'August' too, is a time of working-class escapism, with the journey of the local population to the seaside heralded by an unlikely sample from the 1969 animated children's TV series *Mungo, Mary and Midge*, newsreader Richard Baker's dulcet tones mixing with samples of seagulls that fly to the sound of a drifting flute. The notion of lost time is differently invoked by a sample in 'September' that describes the Calendar Riots of 1752, when the population demanded back the eleven days seemingly taken from it when the British Julian calendar was brought into sync with the European Gregorian calendar (something that also recalls the run-up to the Brexit vote when the UK government argued that leaving Europe would allow for a review of the imperial weights and measures the EU had 'stolen' from Britain).

Viewed from Bonfire Hill, England is a nation where there are clear links between past and present, with both continuity and change registered in a dense sonic collage offering snapshots of a land where things stay the same even as they evolve. Ideas of progress are questioned, with things sometimes considered to be unfashionable or outdated – Morris dancing and folk songs, for example – given equal status with electronic music and sampled loops. And all the time the landscape is to the fore, made strange, sampled, rewound and remixed. *A Year on Bonfire Hill* asks us to listen carefully to the echoes of those who travelled before us, the former occupants of the land whose ways of life have often been obliterated in the march of progress. Explicit here is the idea that in

Figure 7.6 Rebecca Denniff (Bonfire Hill). Courtesy: R. Denniff.

a multicultural society, English people still need to connect with the history of the landscape, 'immersing themselves in the seasons, finding tiny little cultural festivals, whatever they may be when you live'. As Rebecca Denniff (Figure 7.6) tells me:

> I think that we've lost connection with the places where we belong. Because we've homogenised our culture, you can pick someone up in one town and drop them into another town, and we all know it's not right. Where's the character gone? For some people, I guess it's totally fine to be able to go anywhere and it

> all be the same, but I do really believe that if we understand where we've come from, we will have better discretionary powers over where we choose to go. A huge swathe of our own intangible cultural assets are exotic to the people who were born and live in this country. We mock Morris dancing and folk singing and all of that . . . but how can we tolerate other people's cultural sensibilities if we don't even understand our own?

A Year on Bonfire Hill doesn't suggest the truest essence of national culture lies in Morris dancing, folk music or observation of pagan festivals, but asks us to explore the histories of these cultural forms and their connection to the landscape, challenging their co-option by exclusionary nationalists.[31] In this way, it works progressively with ideas of the past, combining established (folk) musical tropes with contemporary electronic sound to create a soundscape that ultimately questions the past, present and future of the English landscape.

Tunes from the Albion Library

Bonfire Hill is not alone in exploring the way the landscape is known through myths and legends handed down through folk traditions, with Folklore Tapes having released nearly a hundred tapes, singles, EPs and LPs on this very theme since 2011. In addition, they have curated site-specific performances, led soundwalks and produced books, calendars and pamphlets on local folklore. Folklore Tapes is not so much a record label, more a collective exploration of music, visual art, history and landscape, overseen by David Chatton-Barker (aka David Orphan) and Ian Humberstone. The former is a fine art graduate with no musical training who began field recording while at Plymouth University. He recalls, 'an important part of living in Devon for me was the immediacy of exploring the surrounding natural world, where I would spend many hours walking the moors, camping and generally absorbing the spirit of the place, noticing the wealth of folklore contained in the surrounding area'. 'Personally', he continues, 'I'm interested in psychic imprints within the landscape and how this can be felt and interpreted, with the historic and individual becoming entwined and the artist a conduit for these energies'.

A chance encounter led Chatton-Barker to Humberstone, an enthusiastic amateur historian dabbling in the DIY arts scene in Exeter. Like Chatton-Barker, Humberstone was driven by a desire to probe the legends that give landscape identity and meaning:

> Folklore is a mysterious and ever-shifting entity, but one bound up with our ideas of identity and heritage. I think it's this combination that makes the subject so appealing. These are tales and customs that belong to us all and help us understand our shared past, but which are often obscure, fragmentary, and neglected by contemporary society. We like to look under the rocks and down into the deep caves of this landscape to see what dwells there.[32]

Their first collaborative release in 2011 set the tone for what followed: a small-run hand-made cassette focusing on witchcraft in Devon. As Humberstone explains,

> we both like to live the physical world, which means working with tangible things wherever possible in our art and music: the two-sided nature of cassettes was also something we wanted to work with, and most of Folklore Tapes' output has been split releases between two artists, each filling out one side of a tape with the overall release unified by a common theme.

The fifty-five cassette copies of *Two Witches* sold quickly, prompting a 10-inch vinyl re-release on Andy Votel and Demdike Stare's label Pre-Cert Home Entertainment. Subsequent releases included a series of 'Devon Folklore Tapes' covering diverse topics including ornithology, ritual and burial, culminating in 2014's *Theo Brown and the Folklore of Dartmoor*, an ambitious set of seven 7-inch vinyl singles of acousmatic soundscapes inspired by folk tales collected by the Devon folklorist Theo Brown, who died in 1993. Accompanied by a hand-stitched book, prints and a thirty-minute film of Dartmoor scenes shot on 8mm film and degraded using organic matter collected from the moors, this release involved sound collage, treated guitar, field recordings, snatches of song and electronic percussion, each track offering a reimagining of one of the folk tales associated with a particular village. Some were full-on psych pop, others more pastorally infused, serene slices of village life. Some are simply downright weird: 'Jan Coo and the Piskies' on the 'Dartmeet' single is pure musique concrète, groaning vocals over ringing chimes and deep drones, referencing the story of the farmhand Jan Coo who heard his name drifting on the wind above Sharp Tor, and who went missing, presumed to have been possessed by the moorland pixies. Described by Humberstone as one of the country's last true wilderness areas, Dartmoor's remote settlements have certainly incubated many bizarre legends of this type, stories folded into its remote trackways, isolated moorland cottages and mossy granite outcrops. Theo Brown documented these fastidiously, with Humberstone and Chatton-Barker retracing her footsteps, from Two Bridges to Widdecombe, Hexworthy to Postbridge, in preparation for this immersive musical project.

The attention to detail in Folklore Tapes' releases and the preoccupation with getting close to the source of local legends, points to the importance of fieldwork as a method. From the start Folklore Tapes projects always involved field trips, with Chatton-Barker describing how a direct connection to landscape, filtered through the photograph and field notes, becomes distilled in recordings. He suggests 'reading can only offer so much', and that it is through fieldtrips that 'unexpected insights come about', before adding that 'on a personal level, I just love being out in the field and this in itself is very positive for both body and mind'.

The label's intensive approach to fieldwork is exemplified in the 2023 release *Brown Wardle Hill*, part of the intermittent 'Lancashire Folk Series'. *Brown Wardle Hill* is based on Chatton-Barker's period as artist-in-residence in the mid-Pennine Spodden Valley, as well as six years living with his partner and child in a cottage in this windswept corner of Lancashire:

> I was fortunate to live on a farm just below the hill itself, and I became aware of a vast amount of folklore relating to the hill and its surrounding slopes. I felt compelled to research and spend time around these sites. There wasn't much recent work into the folklore around Brown Wardle Hill that I could discover and this drove me further into the self-appointed task of unearthing the tales and characters. I found one book by William Robertson [*The Vale of Whitworth, Its Moorlands, Favourite Nooks, Green Lanes, and Scenery*, 1897] which became my main reference book. I explored the museum in Whitworth and further afield in Rochdale, where I found much lore and legend; I also talked with locals in the valley.

This research resulted in both a book (*Lorelines*) and an album with the Whitworth and Healey Vale brass band, recorded live in their rehearsal room near Brown Wardle Hill itself.

The use of a traditional Lancastrian brass band adds verisimilitude to the release, the strident and lively music exuding a pride in place and referencing a rich heritage of brass bands which goes back to the nineteenth century, originally an attempt to ward workers away from the dangers of drink by engaging them a morally productive leisure pursuit. Further connecting the music to the local landscape is the subtle use of field recordings captured by Chatton-Barker at several of the locations referenced on the album: bird song, the wind through the fallen walls of a ruined building, and the babble of water from a spring all contribute texture and tone.

The eleven compositions on *Brown Wardle Hill* are inspired by stories covering a broad timespan from the Neolithic to the present, referencing sites both well-known and long vanished. 'The Famine Tower', for example, refers to a 28 foot high stone folly built by out-of-work mill textile workers during the cotton famine of the nineteenth century; 'The Queen of the Well' speaks of the legend of a spirit residing in the hill's churn well, and the 'Ghost of Intack Farm' references to a reputedly haunted farmhouse, demolished in the 1970s, once home to a man falsely accused of forging banknotes, hung for his crimes at Lancaster Gaol. These stories reveal a rich cast of local characters reanimated by Chatton-Barker, his fellow travellers in a project linking biography and landscape. Chatton-Baker even became involved in archaeological digs in 2020. One object he found on the hill – a brown, broken clay dish of indeterminate age that appears on the cover of the album – has also been occasionally incorporated into live performances as a percussive device.

The sum is a richly rewarding collaborative multimedia project, and testimony to Chatton-Barker's immersive approach to landscape. Like most Folklore Tapes releases, it constitutes an artefactual record of the regional character of a part of England in which the sonification of landscape is central (notably, all but two of Folklore Tapes are about England). As Chatton-Barker details, this multimedia cataloguing is meant to make the landscape 'accessible to an audience who are perhaps more inclined to process information through listening' with the audio a 'way in' for those who might otherwise be uninterested in questions of folklore. Moreover, by archiving the material in different media, Folklore Tapes preserve their landscape research in the form of physical artefacts that they hope will have a life beyond 'the archives'. Chatton-Barker explains:

> At its core Folklore Tapes has around 250 followers who, more-or-less, buy every edition we release, and because of this devotion this audience is amassing a great knowledge of folklore and associated sites. As each new release is shared with the world, so to the audience will begin to expand their knowledge of the landscape and the tales that are associated with it. People can walk to these locations, tell stories and share them with the people who are with them. 'Songlines' are often passed down in families, passing on important knowledge and cultural values . . . it's our hope that the archive we are establishing with Folklore Tapes will be passed down in its own way.

To what extent this archive allows grasp on the vexed question of English identity is something Chatton-Barker has considered at length, concluding that 'Hopefully it's expanding on what people already know and assume about England and its people through a lens into the past, but also through reinterpretation, looking into the present and potential futures'.

Folklore Tapes releases have involved work by a large group of artists and performers, each with their own unique approach to the landscape: Sam McLoughlin, Dean McPhee, Bridget Hayden, Paper Dollhouse, Rob St John, Thorn Wych, Pefkin and Children of Alice (including Broadcast's James Cargill) to name but a few. Some are familiar names in the contemporary folk scene, others adopt a more experimental or wyrd approach. This variety is exemplified in Folklore Tapes' exhaustive Ceremonial County series, which, over the course of twenty-four tapes issued monthly between 2024 and 2026, offers a gazetteer of myths associated with the forty-eight traditional counties of England, some of which have disappeared (e.g. Rutland) or were divided into metropolitan regions (e.g. Lancashire) following the administrative reorganizations of the 1970s. Issued in a limited edition of one-hundred, each with a hand-numbered fragment of an OS map enclosed (Figure 7.7), each tape offers two fifteen-minute tracks addressing local folklore in a particular county, with most of the artists choosing to incorporate a field trip in the compositional and recording process.

Figure 7.7 Ceremonial County Series Tapes I-VI, 2024. Photo: author.

Artists have free reign to create whatever they feel responds to the landscape lore they encounter, with Chatton-Barker revealing he is 'always delighted by the material handed in', noting 'the environment shapes the music massively'.

The Ceremonial County series is nothing but varied: it includes guitarist Dean McPhee exploring the legend of the Bradford Boar (in *West Yorkshire*), David A Jaycock evoking Arthurian legend via repeated acoustic melodies and floating synths (in *Cornwall*), while in *South Yorkshire* Slug Milk use eerie accordion drones and hurdy-gurdy to reference the grisly nineteenth-century scandal of 'Wardsend Cemetery' in Sheffield, where the sexton was accused of digging up corpses to sell them to medical schools. Another stand-out entry in the series – 'In the Shadow of Golden Cap' (*Dorset*) – is by multi-instrumentalist and illustrator Will Yates, recording as Memotone. Beginning with footsteps crunching over a shingle beach, this soundtracks an ascent from the shoreline to a cottage at Morcombelake. The music is an exercise in sparse restraint that builds slowly, with significant nods to the music of Talk Talk, Jon Hessell and Archie Shepp, the whole infused with a late summer haziness. This jazzy confection leads us through a topography of flint and chalk, along paths of tangled bracken and sea buckthorn, with the sound of a babbling brook always in the background. A sax gently serenades over gentle vibes and keyboards as we are taken through a medieval hamlet and back to the ancient well of St Candida, a seventh-century Saxon killed for her faith. The water from the well is celebrated for its healing properties, not least its ability to repair failing eyesight, though it's now green from the phosphorus leached in from surrounding cattle farms. Yates's track is nonetheless a celebration of the stream having brought life to the landscape. As he puts it, 'It has created beautiful little areas for growth and habitat that would otherwise have been grazed or tidied, with lizards, adders, rarer birds and wildflowers, and bigger mammals like hedgehogs and badgers all enjoying the environment and moving with its shifting boundaries'. But the track perhaps gains a poignancy from the fact that this is a landscape whose future is less than certain, with Yates' family having rented a holiday cottage in Morecombelake every summer for thirty-three years until it was taken over by a holiday lets company and became unaffordable. He tells me 'They also modernized it, laminated the living room floor, and we couldn't bear to experience it', a somewhat familiar refrain among those who have been pushed away from the gentrified coast.

The geographical, musical and thematic variation in the Ceremonial Counties series ultimately makes it difficult to pigeonhole it as hauntological, though

Chatton-Barker tells me he has no problem with being lumped in with other practitioners, writers and artists who adopt hauntological approaches. But even if Chatton-Barker feels that describing Folklore Tapes as hauntology is wrong, their work spins out from traditional folk songs and stories to encompass a much wider interest in the role of music in nation-building, past and present. This given, resurrecting and remixing weird tales of fairies and goblins, or sacred wells and haunted villages, becomes a potent means of addressing questions about identity at a time of pronounced anxiety about who the nation belongs to. For the acts and labels discussed in this chapter, folk is invoked not in the hope we can ever return to some lost, pre-industrial England that never was, but is presented as a productive alternative to the lost hopes and failed promises of the present. In different ways, they all look forward to a time when, in the words of George Parr, 'the pastoral alchemy and gothic psychedelia of the island's rich and diverse countryside is no longer a quaint trend subtly running through our traditions and art, but is at the very forefront of our culture'.[33]

Further Reading

Brawn, Jason D. (2023). '"Sounds of our past": The electronic music that links folk horror and hauntology'. In Edgar, Robert & Johnson, Wayne (eds), *The Routledge Companion to Folk Horror*. Routledge, 296–307.

Cornish, Jack (2023). *The Lost Paths: a history of how we walk from here to there.* Michael Joseph.

Holloway, Julian (2022). 'Moving sonic geographies: Realising the eerie countryside in music and sound'. *cultural geographies*, 29(4), 547–563.

Paciorek, Andy (2023). 'Yesterday's memories of tomorrow: Nostalgia, hauntology, and folk horror'. In Edgar, Robert & Johnson, Wayne (eds), *The Routledge Companion to Folk Horror*. Routledge, 183–193.

Rodgers, D. A. (2019). 'Something "wyrd" this way comes: Folklore and British television'. *Folklore*, 130(2), 133–152.

Scovell, Adam (2017). *Folk Horror: Hours dreadful and things strange.* Liverpool University Press.

Simpson, Jaqueline (2009). *Folklore of Sussex*. The History Press.

Thurgill, James (2019). 'A fear of the folk: On topophobia and the horror of rural landscapes'. *Revenant Journal*, 5, 33–56.

8

Deep Time and Psychogeology

On May Day 2023, Yorkshire-based electronic artist and producer Craven Faults debuted a new track, enigmatically entitled 'Parcel 451', as a live YouTube broadcast.[1] The black and white video began with a locked-off video camera, seemingly nestled in the wet grass at the edge of a damp-looking pasture. In the foreground, there are sheep grazing: the fact some have lambs suggests the video was made in the spring, yet the sky is ashen grey, and sleet falls intermittently. In the distance, we see an undulating treeless hilltop ridge, traversed by a dry stone wall. After thirty seconds or so, music begins, a single ominous piano chord which turns into a sustained, low-toned hum, processed through a modular synth. After two minutes, we begin to hear percussive drum and bass notes, regular and slow, atop the swirling synth notes that rise and fall; after ten, arpeggiated sequences can be heard in the layers of sound; around twenty minutes in, another chiming synth joins. The sheep move around the field, mostly in the distance, but occasionally in the foreground: their wool matted and muddy, they show little interest in the camera. At one point, a lamb stands in front of the camera, the landscape framed through its legs. Sounds fade in and out almost imperceptibly as the music swirls, hypnotically.

The stream is watched by a small but obviously rapt audience of several hundred, who exchange thoughts on the live chat. One compares the downtempo, slowly unfolding to Kraftwerk; another mentions Tangerine Dream. One suggests that it feels like nothing is changing until you notice that something has changed, almost imperceptibly. Another points out that Craven Faults are named after the crustal fault line that goes across Yorkshire, where the millstone grit of the Pennines meets the carboniferous limestone of the Dales. They comment that the music is almost tectonic in feel, slow and deep. After an hour, someone asks when the track is going to end, and another speculates that it could go on forever. After a full seventy minutes, some of the layers slowly disappear, leaving the single reverberating chord. As it swirls and subsides, we cut to a shot of the rutted, muddy lane behind the field. The camera, now held head-high, moves slowly through a gate in a dry stone wall towards a large, seemingly abandoned, stone-built agricultural building. As it pans along the wall of the building, it swivels to dwell on a poster, pinned to a wooden stake, announcing the band's forthcoming September 2023 gigs at Thwaite Mill, Leeds. Seventy-five minutes after it started, the music stops and the video fades to black.

The announcement of a gig at this juncture was significant: while Craven Faults debuted with a trilogy of well-received EPs between 2017 and 2019 – *Netherfield Works, Springhead Works and Nunroyd Works* – and released an acclaimed first LP *Erratics and Unconformities* in 2020 on Leaf Records, his identity remained unknown. A 'live' LP released in 2020 was recorded (and filmed) in Craven Faults' studio in an old textile mill in a small town to the North of Leeds, but the videos accompanying this singularly focus on a huge modular synthesizer, spread across three cabinets. These videos show the console flashing with urgent red, green and amber LED lights, hands flicking across switches, twisting dials and adjusting patch cables. But we see nothing of the performer, and record sleeves give no further clues as to his identity. Interviews are rare, the only notable one, in *Electronic Sound* in 2023, hinting at a past in 1980s indie bands. When the debut live performances finally occur, the audience at the Thwaite Mill shows are asked politely not to take photos or record proceedings.

Figure 8.1 Parcel 18, Mallerstang Edge, Yorkshire/Westmorland border. Photo: Craven Faults.

The enigmatic anonymity of Craven Faults inevitably encourages deeper scrutiny of the accompanying record sleeves, which are picked through for clues about the music's production. Impeccably designed (by Yorkshire-based agency Split), these feature large black and white landscape scenes on the cover, images of abandoned mines, dry stone walls, narrow country lanes, former textile mills and tumbledown barns – the iconography of the post-industrial North, in effect, rendered in monochrome. After the first EPs, which featured commissioned photos, all have been taken by Craven Faults, with both the debut album and its follow-up *Standers* (2023) featuring lavish twenty-page photobooks of these images with the vinyl edition, the majority taken under dark skies, gritty and atmospheric. These show landscapes devoid of people, the emphasis on shapes, forms, shadows, texture, with the same depth of focus throughout the frame (Figure 8.1). Craven Falls explains his approach to me:

> I decided to buy a reasonable but cheap secondhand camera. I'm looking for the ordinary in the photography, the juxtaposition of interesting subjects almost to the point of the abstract, where I am looking for shapes in the landscape. I think that removing colour from the process helps, it removes the obvious beauty of a spectacular view and leaves something slightly different. One aspect I try to avoid is making them dramatic. I'm trying to make the subjects look as natural as possible, I want them to be black and white versions of what anyone can see when they are in those places. There's a kind of mundanity to what I'm hoping to achieve. I want them to look accessible and not other-worldly.

As in the work of Fay Godwin or, before her, the Romantic Modernists John Piper and Paul Nash, here photography is a form of archaeology, referencing a landscape's historical layers, bringing them into view and making the past immanent.[2] There is little of recent provenance on display in these scenes, and where there is – a mobile phone mast, for example, or a row of pylons – it gives a sense of scale and rhythm in the landscape rather than accentuating the impact of recent human activity. There are no scenes of tourists thronging chocolate-box-perfect villages nor any Grand Design conversions suggestive of the type of rural gentrification that has increased in pace in rural Yorkshire since the COVID-19 lockdown encouraged working from home.[3]

This obvious interest in landscape is mirrored in the titles of tracks. While some – 'Foddergang', 'Ings', 'Deipker' – appear to be settlement names, others – 'Cuppola Smelt Mill', 'Engine Fields', 'Tenterground' – hint at former industrial activity (a tenterground being an area used for drying clothing in the fulling process). Craven Faults tells me:

> With all the track and album titles I'm essentially looking for something that sounds good to me and evokes some kind of feeling. It can suggest an ambiguity (as in the name Craven Faults), apparent nonsense ('Slack Sley & Temple' – this is actually connected with the spinning industry), a generality (Signal Post, Tenterground), something site-specific (Hangingstones, Doubler Stones, Severals, Weets Gate), industry (lead mining and smelting have given me 'Sun Vein Strings' and 'Cupola Smelt Mill'). 'Foddergang' interested me because I think it sounds intriguing . . . It's actually the name of the small passage found in cow houses (like small barns) to enable the feeding of cattle. 'Deipkier' is a small wooded area not far from my studio: its shape on the map drew my eye at first and the name sounded interesting. On some of the older maps it's called Deepcar and I can imagine a 19th century cartographer asking a local person what the wood is called and naturally being told the answer in the strong local accent. I'm being fanciful here but it's all part of the reason why I chose to use it for a track name.

Track names referring to now-relict traces of the industrial revolution of coal, machinery, steam and textile production hence jostle with those of field and village names which have their origin in Old Norse language, from a time when the only occupants of the windswept moors were those eking out a paltry living by farming marginal land.

This all gives the impression that the music is not so much about specific places but the way the landscape of Northern England has been fashioned from the natural landscape through the layering of successive human activities, ways of life and, importantly, industries. Even the 'Parcel' numbers that serve as catalogue numbers for Craven Faults releases refer to the way the past landscape was carved up and made legible: these 'Parcels' refer to the indexing of nineteenth century 25-inch Ordnance Survey maps which numbered each field and indicated their acreage, meaning that the location photographs were taken in can be identified with some precision (so long as one knows the right OS sheet to look at). This inventive geolocating system also allows Craven Faults to number events, gigs and videos in the same way, in the tradition of the great Northern independent label Factory Records, which used to catalogue artefacts, films and even interviews with a 'Fac' catalogue number: Parcel 428 is a gig in Bristol, a remix EP is Parcel 731, a t-shirt Parcel 272.

Together, the photos and track names are deeply evocative, albeit there sometimes appears a mismatch between them, with the images and record names sometimes lacking any obvious connection. As Craven Faults tells me, this is deliberate:

> I'm trying to create an idea that there are many things to discover in the landscape and I don't want the releases to be linked absolutely to a particular place. There are usually one or two connections but I feel that to make everything fit to a particular album or EP would create a kind of concept album which I'm really not interested in doing. I like to keep things open and free-flowing as this allows for using what I think are the most interesting titles and photographs to complement the music.

All of this suggests a relationship between the music and its conceptual framing that's notional at best, and perhaps wilfully misleading. But probing further, it's clear Craven Faults sees the music, track titles and photographs as working together in some way:

> The titles and images are applied to the tracks at the point of near completion. It's a continuous process of music, research and photography, simultaneously collecting and creating an archive of sounds, photographs and notebooks of ideas. When working on an album I may have 10 or 12 pieces of music on the go, a series of photographs and a collection of names and titles. At this stage of the process I start to see connections between all three and concentrate on which ones appeal to me the most. I don't take a name I like and see which track it would suit best, they just seem to come together quite naturally.

Though Craven Faults incorporates some field recordings in his work, these are mainly used as a source of rhythmical sounds, their presence not much helping to locate the music in time or space: as with much of the music inspired by landscape, without the presence of track titles and sleeve images, the fact that this is about landscape at all would perhaps not be evident to the casual listener. But when combined with the graphic design, photography and track titles, the beatless, modular loops and drones begin to make sense as an updating of the 1970s *kosmiche* krautrock of Cluster, Harmonia and Kraftwerk, transplanting this from Germany to Yorkshire to forge a machine music summoning a history of land and industry:

> The link between the music and the landscape can seem tenuous . . . I do like to think the music is influenced by the landscape but I'm unable to pinpoint any definite instances of this. I don't take music into the landscape, I never listen to music whilst out walking but I will think of ideas for new pieces when I am outside. I am possibly not the best person to make connections between the

> music and the landscape because I am so closely tied up with its composition in all its technical aspects as well as its purely creative ones. The listener is probably in a far better position to make connections than I am!

Indeed, there is little about 'Cuppola Smelt Mill', a nine-minute track named after a mill built in 1792 on Grassington Moor, that necessarily evokes the reverberatory furnaces used to smelt lead. But the regular electronic throb that underpins the track was suggested by some reviewers as alluding to industrial processes. This is not industrial music in the mould of Throbbing Gristle or Nurse With Wound, nor the electro-industrial rhythms of The Human League and Cabaret Voltaire forged in South Yorkshire in the early 1980s, but a more meditative contemplation of the nature of industry and the scars it leaves in the landscape. Likewise, on 'Slack Slay and Temple', a metronomic pulse is joined by a slow, percussive bass that drives the track for eighteen long minutes, a rumbling hypnotic drone one reviewer described as the sound of a 'dark satanic mill': the mill machinery heard briefly at the start of the track is actually a recent recording made in the (now-closed) factory producing balls of wool, adjacent to Craven Faults' current studio.

The location chosen for Craven Faults' first (sell-out) gigs in September 2023, Thwaite Mills, hence seems more than just a little serendipitous (Figure 8.2). Set between the River Aire and the Aire & Calder navigation canal to the

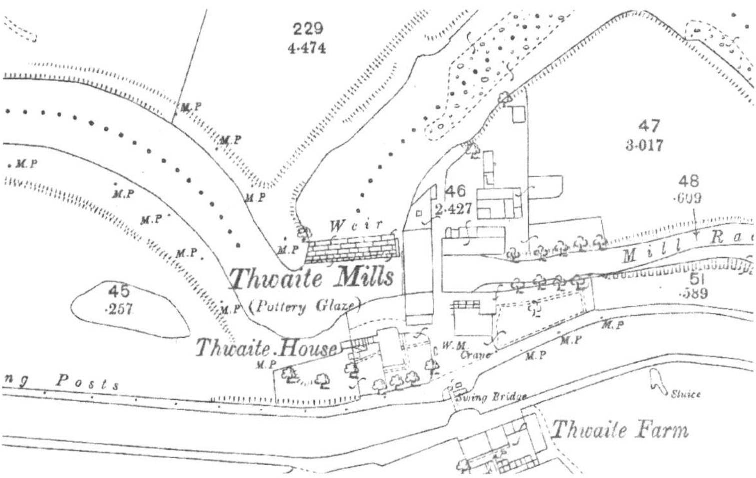

Figure 8.2 Parcel 46, OS Map, surveyed 1891, published 1893. Courtesy: National Library of Scotland.

south-east of Leeds city centre, Thwaite Mills is an industrial museum with functioning waterwheels. In the past, these were used in various industrial processes: initially, the manufacture of woollen cloth, and later grinding chalk to make whitening for paint. In the wordless Thwaite Mills programme (Parcel 46), monochromatic images of the mill's waterwheel and industrial machinery are juxtaposed with close-ups of the tangled cables of Craven Fault's modular cabinet, dials labelled with gaffer tape. One image shows the circular Scotch real-to-real tape used to create tape loops on 'May Birching' immediately below a photograph of the waterwheel and the pulleys and gears linking it to the industrial machinery (Figure 8.3). This also recalls the labelling of the first three EPs as 'Works': here 'work' has a double meaning, both a site of production – a factory – and the work of art itself. When we talk at his studio, he tells me that he despairs at the lack of craftsmanship displayed in much contemporary music and bemoans the fact that generally people seem to want stuff made cheaply in China rather than using the skills that are so often under-utilized in the UK. Though he concedes that music can sometimes come together suddenly, he also suggests he might work on a track for two years or longer, revisiting old edits and trying to make things work: he laughs that the 'cutting room floor' is full of things that sound fantastic but are just 'not right'. The suggestion here appears to be that the production of music on a modular synth in an old factory is a form of craft production equivalent to that of master weavers at the loom, with Thwaite Mills the perfect backdrop for a music taking inspiration from a landscape of industry and toil.

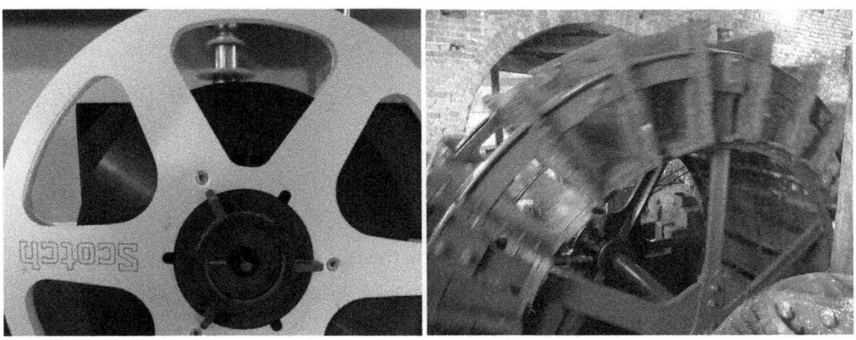

Figure 8.3 A tape reel in Craven Faults' studio (photo: author) and Thwaite Mills' waterwheel (courtesy: Wikipedia Commons).

A Northern landscape?

Much of the music of Craven Faults refers, explicitly or otherwise, to the long-standing connections between land and industry in Yorkshire (and occasionally neighbouring Cumbria) and the way these continue to inform the post-industrial present. In this sense, it plays into myths about Northern identity which have been imposed on the region by a Southern cultural elite, and which, in turn, have been perpetuated by those Northern folk who have readily bought into a mythology of the North as altogether wilder, more industrious and harder than the soft, pastoral South. Leeds-based sociologist Karl Spracklen suggests there are still many Northerners who embrace these myths by arguing for the maintenance of an 'authentic' working-class culture, custom and language that has been near-wiped out by decades of de-industrialization.[4] But there appears a world of difference between the mythos spun by 'professional Northerners' like Stuart Maconie or Ian McMillan and Craven Faults' almost-phenomenological reading of landscape history. This is one in which identity is understood to be shaped by an encounter with the materiality of the landscape, particularly as experienced through *journeying*. This is a metaphor that occurs frequently in discussions of Craven Faults' work, not least because Leaf Records described his first album as 'half-remembered journeys across post-industrial Yorkshire'. But Craven Faults tells me, for him, these journeys actually start with a '20-minute road trip' to his studio through country roads with views across the moors and into the Yorkshire Dales National Park, the sunlight catching the top of Baildon moor, above Menston village, the site of numerous Bronze Age carved rocks. 'When the light isn't on it', he tells me, 'it just looks like a normal hill, but then at certain times you can really see the rocks underneath the earth, and it looks different to any surrounding peaks, magical, in a way'. As concedes, his music could have been about almost anything, but these journeys were the direct spur for his project:

> This journey and the studio location have been an inspiration in as much as I am always aware of the landscape around me: industry, mills, former mills in the town, neolithic sites, old dry stone walls on the hills . . . Travelling through places is all part of the concept, both the journey and the destination. There is also the journey taken along footpaths and being aware of what is there to see, not just hills and valleys and the other aspects of 'beautiful countryside' but also, and probably more so, the minutiae and smaller details, the old broken gate post, the clapper bridge over a beck, the abandoned barn or cow house.

This eye for detail suggests an acute awareness of the way that landscapes are made and an unlikely adherence to the principles of landscape history as spelt out by one of its founding fathers, W. G. Hoskins, who argued, 'the English landscape is the richest historical record we possess, and one must observe the smallest details'. The implication of Hoskins' best-known work – *The Making of the English Landscape* – was that it was not necessary to be an archaeologist to read the material traces of the past.[5]

Craven Faults' project came into focus when he realized his instrumental music was 'a product of my character, or certainly of aspects of myself' moulded by the landscape. While having no expertise in industrial heritage, nature or archaeology, he suggests he has always held 'a deep curiosity' about his surroundings:

> As an example, I often wondered why a field of green grass could be separated from rough moorland by nothing more than a wall. Sometimes the field would be surrounded on all sides by moorland. I learned that the field was called an Intake, land taken in from the moor, 'sweetened' with lime to reduce the acid content of the soil and used for grazing sheep. As a child I remember there was an Intake Lane not far from my grandmother's house in Bradford. I liked the sound and the idea of the Intake and so it became the title of a track on the second EP release (*Springhead Works*). The lime used on the land led me on to finding out the small arched stone constructions scattered all over the dales are lime kilns built to provide lime for the farmers. One thing often leads to another when looking into the history of the human intervention in the landscape . . . I find it interesting to plot changes over time and also to see what has remained the same, such as the routes of roads, tracks and paths that were originally rural and are now completely absorbed into the urban landscape but retain their original courses.

Erratics and Unconformities has a dedication to Arthur Raistrick, whose 1972 book *Industrial Archaeology* started Craven Faults off 'on this journey' when he purchased it in the 1980s. When I meet Craven Faults in his studio, he brings the book along, alongside several others, and we note the coincidence between the cover design of Raistrick and Illingworth's (1949) *The Face of North West Yorkshire*, and Craven Faults releases, notably the *Netherfield Works* EP, suggesting a shared way of looking at landscape. Raistrick was a Yorkshire polymath who became an expert in Chinese porcelain and the music of Bach but, alongside W. G. Hoskins, is best known as a pioneer in English landscape studies, producing over 300 books, pamphlets and articles, mainly on the landscapes of the North of England. But unlike Hoskins, who argued little of any beauty had been created after the eighteenth century and searched for quiet solitude in the fields of medieval Leicestershire, Raistrick argued the quintessential beauty of the Yorkshire Dales – dry stone walls, vernacular buildings and packhorse ways – was almost entirely the product of the industrial era.[6]

Raistrick was a pioneer of industrial archaeology, but he is also notable in landscape studies for bringing a detailed understanding of geological processes to the fore in the understanding of regional landscapes, having been appointed as Lecturer in Geology at Durham University in 1929. He related the building of 'miles and miles' of dry stone walls in the North Pennines to the glacial processes that shattered the carboniferous limestone, creating fragments that can now be readily found in fields, and assembled without mortar into walls that typically outlive masonry, important in windswept areas where hedges would not flourish either.[7] The familiar pattern and texture of the contemporary Pennine landscape is then, he explained, explicable only in the context of the geological past.

In Craven Faults' music geology is also given due prominence as shaper of the landscape, both physically and culturally. The mini-album *Enclosures* (2020), for example, begins with 'Doubler Stones', referencing the two near-identical mushroom-shaped rocks atop Ilkley moor, their caps blue-hued gritstone, softer sandstone below. Millstone Grit is, as Ben Myers has noted, as perfect a metaphor for Northern England as can be found, being stoic, solid, immovable, ancient.[8] Glyn Hughes famously described the Pennines as a slab of millstone grit that occasionally breaks through the surface like 'bones through the flesh of a hungry cow'.[9] This bedrock is alluded to in much of Craven Faults music by dint of duration and the layering of sounds which build slowly, but on 'Doubler Stones' the spirit of Tangerine Dream is invoked via a track that suggests a drive through an unfolding landscape, with periodic outcrops of electronic noise overlaying the motorik beat. The second track on *Enclosures*, 'Hard Level Force', refers to a waterfall above Old Gang Smelting Mill (depicted on the cover), while the final track – the seventeen-minute shimmering epic 'Weets Gate' – refers to a boundary stone marker on the route between Malham and Fountains Abbey.

On *Enclosures*, stone appears in multiple forms: as rock eroded by wind and water, hewn by the elements and assembled into dry stone walls, cairns and boundaries (Figure 8.4). Here, *Enclosures* has a double meaning, both in the sense that people have made long made enclosures from stone, marking out their territory in the landscape, but also that the privatization of the landscape was driven by parliamentary Enclosure Acts, with dry stone walls proliferating as the

Figure 8.4 Craven Faults' *Enclosures* Courtesy: Craven Faults/Split Design/Leaf Label.

commons was carved up into private parcels of land. The second Craven Faults album, *Standers* (2023), also alludes to the way stones have been used, with the title referring both to the stone supports used to hold up props in mine shafts, as well as rocks standing proud in the landscape. Opening track 'Hurrocstanes' refers to boundary markers (a 'hurroc' or hill of stones), the name corrupted into that of a hamlet, Haddock Stones, near Otley. The cover of *Standers* depicts Nine Standards Rigg, dry stone cairns on the border between Cumbria and North Yorkshire whose purpose remains conjectural. Having stood for at least 800 years, one theory is these were erected by the Romans to appear as soldiers standing on the horizon. A more prosaic theory is that they represent a territorial divide in the landscape.

These features in the landscape have also fascinated Richard Skelton, a self-styled artist of Northern landscapes. A prolific writer, recording artist and eco-poet, Skelton makes his modern classical and experimental music in situ as a response to the environment. The West Pennine moors of Lancashire and South

Cumbria have been one of his inspirations. Having moved to a small farmhouse there following the death of his wife in 2004, his work began to explore questions of provenance and location by forging a sometimes-visceral connection with the land (one performance involved him exhuming and playing a buried violin, another playing a guitar with a piece of bark as a plectrum). He has also completed a PhD in placemaking at Manchester Metropolitan University (on paleolithic responses to landscape change in Northern Britain), writing that music, like creative writing, can encourage a sensitivity to the non-human agency of landscapes and the historical *strata* that compose them. In his words, 'Coming to terms with what is before our eyes inevitably requires a comprehension of what has gone before, historical, archaeological, ecological, geological. To look at the present is to look into the past'.[10]

In this spirit, *Beyond the Fell Wall*, a film and book project completed in 2015, considers the presence of stone walls in the uplands of Cumbria, making these come alive as agents of landscape change. In the twelve-minute video, static black and white shots of stones, scree, rivers, tree trunks and fields are accompanied by electronic drones made by Skelton to represent the slow, natural forces of weathering. Walls are traced back through time – rubble field, outcrop, sediment, magma – with the suggestion they have been made over a timespan that evades comprehension. But as with Craven Faults' interest in place etymology, Skelton also explores the vocabularies through which walls become known, providing a 'lithic' glossary of terms familiar to dry stone wall-makers past and present: cams, cappers, coping. And, like Craven Faults, Skelton has an interest in the way that stone walls outlive their makers, with nature reclaiming them, buckling and bursting them over time. His is a history of dereliction, emphasizing the precarity of human life in the exposed, elemental landscapes of the North. Invoking ideas of deep, geological time, he describes how the stone outlives the wall, asking how long a wall can survive on the ground, on a map and in the imagination.

In *Underland*, Robert Macfarlane argues that focusing on the way geology manifests in the landscape is not a means of escaping the troubled present but reimagining it.[11] Here, he invokes hauntology by suggesting that climate change is making us aware of 'what lies beneath' in multiple ways (e.g. retreating glaciers exposing rock, drought revealing crop marks of former settlements, preserved skeletons being exhumed in melting permafrost). Faced with life in the Anthropocene, and the extinction crises it has precipitated, Macfarlane offers a prescription based on an awareness of deep time which sees human life as both a 'gift and inheritance' stretching millions of years into the past, and millions more in the future. Engaging with the horizons of deep time becomes a way of encouraging a more ethical relationship with the environment than one based on blasting mountains and extracting the earth's minerals and fossil fuels in the name of profit. While any piece of music, like writing, is ephemeral in the wider horizon of geological time, in the context of planetary destruction, it reveals how temporalities of different depth and duration haunt the present.[12]

Both Richard Skelton and Craven Faults express fascination with a past whose duration goes back to a time before human occupation, reframing ideas of hauntology to consider not just the lost futures of the human past but the lost futures of the *planet*. In many ways, this music is about national and regional identity, but it considers the Northern landscape in a much wider, and longer, context. In the words of Craven Faults:

> I certainly don't want to put across any ideas of Englishness or of being British. The geography/geology takes you back to a time before human occupation, to a time when the Yorkshire Dales were near the equator. The history is about movements of people. Both these subjects are about the very slow but continual shifts over time which are still ongoing today. For me, the past in the landscape is the discovery of elements that have remained over millennia, even those that are barely observable. *Erratics & Unconformities* refers to geological action. Where does geology stop and history and geography begin?

Craven Faults' journeys, guided by the underlying geology below as much as the human landscape above, can hence be considered a veritable *psychogeology*, a transect of the landscape guided by the geo-psychic contours of what lies beneath: limestone, grit, sandstone, and slate, rich lead veins and coal seams, and cavernous gills, pots and caves.

Psychogeological perumbulations

Recording as Spaceship, Mark S. Williamson is a primary school teacher, part-time megalith hunter and musician. Inspired to explore the landscape by Julian Cope's *Modern Antiquarian*, he spent four summers working near Carnac in Brittany, site of one of the largest alignments of menhirs in Europe, and regularly contributed to Cope's *The Megalithic European* project (published in 2004). Encouraged by Cope, he recorded drone rock, but slowly graduated to more field-based and site-specific musical production. *A Prospect of Loughton's Brook* (2017), for example, traces a route through Essex from Epping Forest to the River Roding, using binaural field recordings augmented with synths, piano and strings. Asked to make an album for Where It's at is Where You Are records in 2018, he found inspiration in the more rugged landscape surrounding his new home in Todmorden, Yorkshire after he moved there from Essex:

> I studied geology at A level and at University and halfway through I realised that I didn't want to work for the petrochemical industry. I knew that this degree was never actually going to take me anywhere professionally. But I really enjoyed it. I just found it really fascinating. I guess it was about . . . the deep connection to the landscape and a deep understanding of where you are. Todmorden is somewhere that actually has bedrock above sea level that outcrops, while obviously down in Essex there is no rock anywhere. I started thinking about geology again and I was teaching it a bit of it to the kids at school because you can kind of do what you want.

In his own words, this reignited his enthusiasm for geology and set in motion a musical project based on the geomorphology and geology of the local area: 'I started investigating where I was and I was having these ideas that I could go to travel the country and go to somewhere where everything's volcanic and go somewhere where it's limestone . . . but then I just thought, we can do this here, within walking distance of my house'.

Here, the increasing portability of music technology allowed him to go directly into the field from his Todmorden home:

> There was a little four-track facility on the recorder that I had so I could go out into the field with music-making equipment, and build-up little pieces. Literally sat wherever it was that I was trying to make music about. And I felt that that would sort of automatically imbue that music with sense of that place. So I first did an album, mainly acoustic instruments . . . droning things, but doing it with harmoniums, carrying all this stuff over fields. But my job had given me an iPad, which I've never had before, and I realised that you could get all these apps that made these incredible sounds that actually replicated the synthesisers that I had coveted as a kid, but they were free . . . I can take an iPad with me somewhere and start adding those sounds into these soundscapes of natural things.

Layered drones on the iPad, when played through a portable speaker, provided the bedrock of his recordings, with Williamson adding midi keyboard effects, woodwind, chime bars and harmonium in the field, capturing the ambient environment as the piece is constructed.[13]

A Transect of Shackleton Knoll is a single seventy-seven-minute track, based on run down the Calder Valley that Williamson had been regularly taking, past a ruined farmhouse, Shackleton Knoll itself, an overgrown Victorian landscaped waterfall valley, a section of woodland river and finally a railway viaduct. This *Transect*, in the classical geographical tradition, follows the twists and turns of the landscape, stopping at those landscape features that resonate with vibrant materiality, recognizing that every landscape starts with its geology. In Williamson's words, 'Each layer, at the surface or below ground, contributes to the appearance, the fertility, the habitability, the safety (or danger) of a place. So for a deep survey of the valley I had to start with the geology'.

Enacting a drift through the layers of time as well as space, this well-received album presaged a trilogy of albums that have developed these techniques of field recording and in situ performance. *Outcrops* (2019) (Figure 8.5), *Ravines* (2022)

Figure 8.5 Spaceship's *Outcrops*. Courtesy: Mark Williamson/Maxim Peter Griffin.

and *Structures* (2024) offer a sequential engagement with the environment that starts (in *Outcrops*) with the raw stone eroded and fashioned by the elements before it becomes building material:

> Exploring the area I found the four outcrops I recorded at. They're all the same Millstone Grit, but the nature of their settings allowed me to 'fit' them to my idea of telling the story of the geological history of the valley. I started at Whirlaw, and that had to be the Millstone Grit. Orchan Rocks is a bit lower down the valley side and to get there from Whirlaw you're traversing the softer mudstones that come between the layers of grit. Once you're back up at Hawk Stones, you're looking over the valley to where the coal measures outcrop. It made sense to focus on them from there. Finally, Bride Stones had to be the last piece as it was formed by the glacier which the track attempts to invoke. So the locations were chosen as a practical route as I worked my way round, over a period of months, making the recordings.

Williamson's method (Figure 8.6) involved finding gaps between rocks that he could literally crawl into, recording there with his keyboard and iPad with no real plan except sensing the stone's history, exploring the acoustic qualities of rock and discovering how it resonated as music was played. The four tracks are lengthy instrumentals, hypnotic experimental synthesizer drones. As he tells me, 'This is millstone grit. This was formed in this big sedimentary basin somewhere 330,000,000 years ago. What does loads of organic matter that been crushed for

Figure 8.6 Mark Williamson. Courtesy: M. Williamson.

millions of years feel like? This is geological time, this has just got to move really, really slowly'.

On the second LP, *Ravines*, the focus moves to the ways rock is eroded over long periods of time by streams and rivers, and sometimes sudden torrents of water, the music reflecting just 'how very violent those places can get'. In comparison to the serene ambient drones on the first album, here things become noisier, the sound of gently burbling water recorded with hydrophones giving way to layered synth drones that evoke the flow of water downstream through steep-sided cloughs, descending into rumbling rapids where stones bounce and split ('Jumble Hole Clough' evokes a roaring vortex of water). On the third album, *Structures*, Spaceship's attention moves to questions of inhabitation, and how the stone shaped the human landscape:

> I started noticing all this human stuff, how this millstone grit was used by the people who came to live here and how that inhabitation evolved. The Millstone Grit and the landscape we get leads to cheap farming, which at least gives you wool, which leads to mills and so on and so forth. And then by the time cotton comes in, the infrastructure is there, because you've already been spinning wool for hundreds of years, the skill set is there. You're just applying it to a different material. Where your underlining bedrock is different, your industry grows different, the way of life is different, and it shapes the way people are.

This final album in the trilogy alights on sites where the links between culture and geology are writ large. 'Stuap's Mill' was recorded at an eighteenth-century cotton mill standing at the top of Jumble Hole Clough, abandoned after a dam collapse in 1896: here additional trombone from Sophie Cooper was used to make connections to the brass band traditions associated with Northern towns.[14] Another track is entitled 'Shore Baptist Chapel', one of many non-conformist churches in Todmorden, founded in the eighteenth century by followers of Dan Taylor, a missionary teacher from Hebden Bridge. Here, Williamson got Sam McLoughlin to set up a folding preacher's harmonium, 'a little pump organ thing', inside the chapel, adding a dimension to the recording by emphasizing the links between the North of England and the non-conformism that seemingly offered more to the local working-classes than the Southern-based Church of England ever could.

These three albums, together with the accompanying film *Strata*, suggests a chronology: the first, dealing with geological time, is typified by slow drones, the latter albums more 'musical' as the influence of human activity on the landscape comes to the fore. Spaceship hence traces the way the 'geology affects the land-use and then the land use affects the psyche, you know, affects the people', Williamson confessing 'it'ss almost a little bit evangelical' to place emphasis on the sometimes unconscious role of geology in shaping society. The notion that he is pursuing *psychogeology* is one that sits comfortably with him:

> When I started doing this, eight-nine years ago, there was a lot of talk of psychogeography . . . liminal edgelands and all this kind of stuff was very much the thing. And I was like, well, I'm not really doing that because my stuff was aligned with geology, but it was also aligned with place and also population and people and everything else. So it worked as a term and it also if I'm brutally honest, felt like something that was a little bit eye-catching . . . So psychogeology it was.

This notion resonates with some of the concepts developed in critical social sciences that help us see how human lives are entwined with the wider biographies of earth systems. For instance, Palsson and Swanson have talked of 'environmental geosocialities' to draw attention to the way that ideas of both historical time and geographical scale need to be rethought in the Anthropocene, the suggestion being that human activity intervenes in and shapes the formation of geological stratum.[15] Similarly, Nigel Clark writes, 'rather than simply affirming that the Anthropocene thesis has suddenly propelled geological strata into political significance' a focus on the entwining of the social and geophysical world reminds us that the geological has 'long played a constitutive role in social and political formation'.[16] For Clark, it's vital that contemporary politics considers what lies beneath, the relation between the exposed surfaces 'to which we can make some difference' and those deeper forces or formations that 'stubbornly cleave to their own agenda' problematized by considering the histories of extraction, disposal and contamination that have disturbed the earth's strata.

But beyond the rugged uplands of Northern Britain, this new-found preoccupation with the way geology shapes landscape and psyche is yet to register in music in any significant way. There are, as yet, no performers writing about silica sand extraction in the Hampshire Basin, no heartfelt paeans to London Clay. Even the chalk that defines the landscape of the South Downs, so central to imaginations of Englishness, is seldom considered in music despite its history as a resource that has been exploited for lime, ballast, cement and bricks. But when acknowledging the importance of 'deep time' in the making of the Northern English landscape, Richard Skelton, Craven Faults and Mark Williamson (alongside fellow travellers including Autumn Richardson, Conflux Coldwell, Amy Cutler and Drew Mulholland) stress the importance of both the land above ground and the geology below it, affirming occultist Ithell Colquhoun's argument in her book *The Living Stones*:

> The life of a region depends ultimately on its geologic substratum for this sets up a chain reaction which passes . . . through its streams and wells, its vegetation and the animal life that feeds on this, and finally through the type of human attracted to live there. In a profound sense, the structure of its rocks gives rise to the psychic life of the land: granite, serpentine, slate, sandstone, limestone, chalk and the rest each have their special personality dependant on the age in which they were laid down, each being co-existent with a special phase of the earth-spirit's manifestation.[17]

Here there is no claim that the materiality of the physical landscape *determines* the 'psychic life of the land', no fetishizing of the geological. But there is an invitation to consider how the present is shaped by complex histories of resource extraction that have bound people to the land in particular ways, creating 'localised specificities of socio-material relationalities'.[18] Exploring these psychogeologies provides a fuller grasp of the forces that have shaped our landscape and might ultimately help us exorcise the spectres of the resource exploitation that has led us to the current climate emergency.

Further reading

Bosworth, Kai (2024). 'Filling the hole? On new geographies of the subsurface'. *Progress in Human Geography*, 48(2), 151–171.

Chivers, Tom (2021). *London Clay: Journeys in the deep city*. Random House.

Clark, Nigel (2017). 'Politics of strata'. *Theory, Culture & Society*, 34(2–3), 211–231.

Colquhoun, Ithell (2017). *The Living Stones: Cornwall*. Peter Owen.

Gordon, Helen (2022). *Notes from Deep Time*. Profile Books.

Irvine, Richard (2020). *An anthropology of Deep Time: Geological temporality and social life*. Cambridge University Press.

Macfarlane, Robert (2020). *Underland: A deep time journey*. Penguin.

Myers, Ben (2018). *Under the Rock: Stones carved from the land*. Elliot and Thompson.

Overend, David, Lorimer, Jamie, & Schreve, Danielle (2020). 'The bones beneath the streets: Drifting through London's Quaternary'. *cultural geographies*, 27(3), 453–475.
Trower, Shelley (2015). *Rocks of Nation*. Manchester University Press.
Wilson, Louise (2023). 'Above and below: The sounding of haunted geologies'. *Moving Image Review & Art Journal*, 12(2), 219–238.

9

The Frayed Edge

Whether the soaring headlands of Cornwall, the sandy expanses of the North Northumbrian dunes or the 'Jurassic' coast of Dorset, the English coastline is a twenty-first-century magnet for recreational walkers, wild swimmers, sailors, bodyboarders, fishermen and sunbathers – at least when the English weather relents sufficiently. But beyond those who actively consume the coast, there are strollers and flâneurs who visit the coast simply to drink up the ambience, among them the many writers, visual artists and musicians whose work is inspired by being by the sea. Quite why the coast continues to exercise such fascination for creative types is open to question, but in some recent writing, it's asserted that being at the coast gives privileged insight into questions of English identity and belonging, the ocular pleasures of staring out to sea entwining with all sorts of questions about the future and past of our 'island nation'. Hester Blum, for example, suggests the sea, rather than land, is now the 'critical location from which to investigate questions of affiliation, citizenship, economic exchange, mobility, rights, and sovereignty'.[1]

It was not always like this, of course. For much of English national history, the coast was marginal to ideas of nationhood. It was the countryside that defined the essence of English culture, with quaint hamlets nestling in rolling downlands, church spires and village greens providing the inspiration for most artistic articulations of national identity. The oceans, in contrast, were culture's Other, hostile, untamed and threatening: a gothic sea that sometimes raged tempestuously, the site of real and imagined terrors. But this conception of the coast changed into the eighteenth and nineteenth centuries, with the sea gradually 'tamed' via its incorporation into rituals of bathing and restorative salt-water treatments. Pioneering resorts like Brighton, Margate and Southend prospered because of their proximity to London, with sea bathing becoming part of the season of balls and festivities enjoyed by the fashionable and marriageable young: these resorts aped the architecture of inland spa towns, boasting Georgian parades, grand hotels and well-appointed reading rooms. From these beginnings, 'beach life' was born, and ideas of the coastal resort as a site for escapism began, with working beaches of boats and inshore fisheries replaced by an infrastructure of rest and recuperation.

Following the Bank Holiday Act 1871, the pleasures of the beach became available to all, with railway and coach trips transporting working-class urbanites to the seaside every August to enjoy what became the familiar staples of bucket and spade tourism: kiss-me-quick hats, Punch and Judy shows, fish and chips, fairgrounds and donkey rides. But into the late twentieth century, this combination of seaside pleasures transformed again in the face of competition from a globalizing tourist trade and the lure of Mediterranean package tourism. As Rob Shields details in his account of the liminality of the seaside, the transgressive quality of the coast, poised between land and sea, began to attract forms of criminality and violence, most famously the 'mods and rockers' conflicts of the 1960s.[2] Images of the seaside as sanctuary were replaced by something less salubrious, with many resorts becoming known more for dirty weekends and alcoholic excess than healthified leisure. Some declined precipitously, their formerly grand hotels and guest houses becoming temporary accommodations for the homeless and refugees.

A 2019 Parliamentary Select Committee on Regenerating Seaside Towns and Communities noted 'a litany of regret and a paucity of ambition' in the fact many English resorts had not adapted to the decline of bucket and spade tourism: while it was noted Brighton and Bournemouth have become creative centres anchored by universities, seaside towns in general have lower than average earnings, high proportions of adults claiming benefits and more seasonal unemployment than elsewhere in England.[3] Even in the affluent south of England, resorts including Bognor, Margate and Weston-super-mare fall into the lowest deciles for mental health, education, drug and alcohol problems. These constitute a precarious periphery, one where cultural disorientation and social stigma accentuate social divides, sometimes fracturing working-class communities along racial lines. As Daniel Burdsey has powerfully argued, 'Coastal spaces have been diverse and inclusive during distinct periods, historically and globally, but have then taken on, or been preceded by, a contrasting role within emergent politics and practices of racial segregation', something that has been increasingly evident in the post-Brexit era.[4]

These faded 'Kiss Me Quick' resorts hence provide rich pickings for hauntologists and psychogeographers keen to divine the essence of Englishness.

Tom Sykes and Louis Natter's (2022) *Coast of Teeth*, for example, charts a seaside landscape of sewage-filled seas lapping beaches where post-lockdown ravers come to violent blows with one another after all-day visits to Wetherspoons pubs.[5] Likewise, Tom Bolton's (2018) *Low Country: Brexit on the Essex Coast* is a hallucinatory walk around the fringes of Essex, rushing through the caravan parks and shacks of Jaywick, the UK's most deprived community, described as rows of chalets that 'seemed to have landed on the pebbles complete with fenced front gardens, served by unadopted roads and surrounded by scattered rubbish heaps piled with mattresses, waiting for a match', before leaving hurriedly for Clacton, likewise described as 'a blot on the landscape' and home (at the time) to the office of the only UKIP MP in the country, Douglas Carswell.[6] Madeline Bunting's (2023) *The Seaside: England's Love Affair* likewise reports 'entrenched deprivation' alongside a 'felt sense of Englishness' and 'toxic nostalgia' in many of the places she visits in her tour of forty British resort towns, and sees in them 'the wider national failures of a brutally inadequate welfare state, growing inequality, failing public services and a crisis of affordable housing'.[7]

Musically, Junkboy is one act that also seems more than just a little preoccupied with the coast. A collaboration between brothers Mik and Rich Hanscomb, their rich, electrified chamber pop is imbued with a rich sense of place. Rich tells me their 2023 album *Littoral States* is shaped both by the mythology of the Sussex coast and field recordings made along the coast from West to East. Some of these recordings are used explicitly in the album, others 'digitally refracted and mixed in to provide a kind of aural bedrock'. Rich suggests the album is partly a response to the death of his father in 2020, and partly an attempt to search for beauty among the palpable decline evident along parts of the Sussex coast:

> *Littoral States* traces a blood line from Bognor (where our dad was born) to Worthing (where Mik lives) into Brighton and out in Seaford (where I live). There was no intention to evoke Englishness but we are celebrating a life and familial links by venerating the environment we have in common. At the heart of our music is 'the seaside' – an English invention no less – and all the contradictions this suggests. The seaside can often be a liminal space for us to pin our own aspirations and anxieties on, a place that offers freedom and temporary respite but is also stubborn conservatism (with a small and upper-case C) manifesting in recent times as petty nationalism. Some of us pushed to the edges seek the opportunity to create and be inspired by the sublime and some of us suffer displacement. I think all of this suffuses Junkboy's musical consciousness like water ingress or salt air corrosion.

Junkboy also suggest that the cultural infrastructure in some seaside towns is far from conducive to experimental music-making. As Rich tells me, 'We also perceive a sense of detachment from the rest of the country living in seaside towns – we certainly feel that we've always been adrift from the musical cognoscenti'.

Dave Clarkson's musical responses to the coastal landscapes of England also hint at this sense of marginality (Figure 9.1). Clarkson, who has previously recorded as Central Processing Unit and collaborated with Vini Reilly of Durutti Column and Graham Massey from 808 State in the 1980s, joined Todmorden-

Figure 9.1 Dave Clarkson, Talacre Beach. Courtesy D.Clarkson.

based kosmische band Lines of Silence in 2024. In between, he released an impressively eclectic series of field recording-based electronic albums centred on excursions to some of the obscure corners of the country, each accompanied by explanatory booklets and photographs. Among these were *Shore Ghosts and Sea Mystery* (2017), *Coastal Quicksands* (2017) and *Music for Lighthouses* (2014), all exploring the ebb and flow of the sea and its ability to exhume memories with every fresh tide. Clarkson's (2022) *A Pocket Guide to Dreamland: Faded Fiaryground and Ghost Towns of the British Isles,* however, tackles questions of coastal decline head-on, using field recordings from seven towns – Margate (where the eponymous 'Dreamland' theme park is located), Rhyl, Blackpool, Hastings, Porthcawl, Rhyl and Whitley Bay – all commonly identified as faded resorts. These recordings, of electronic arcade machines, penny falls, fairground organs, rollercoasters and even his own family's visits to the seaside, were remixed, spliced and sliced into tracks that summon forth the ghosts of days gone by:

> My musical influences here are generally artists who are willing to work outside their comfort zone; sound recordists, musique concrète, sound collages, also music which has a strong sense of melody and is executed in a very simple way such as traditional British folk music or even Kraftwerk. But the main influences on my *Pocket Guide* albums are actually non-musical: old travel guides from the 1950s-1970s and Ordnance Survey maps of areas I visited when I was young. Like these, landscape recordings are a snapshot in time. The ambience of the locations is ever-changing due to development, weather and other factors, and I am interested in field recordings as ingredients for sound collages or musical compositions. It's like painting with a palette of 'location recording' colours.

Here, Clarkson directly references coastal change, and the contrasts that are offered by seasonal shifts, with seaside resorts in mid-winter appearing grey and inhospitable as the winds whip in and the tides crash over the seawall, as opposed to the candy stripe kitsch of midsummer. As Clarkson notes 'I find they can either be calming locations or intense sound experiences . . . a coastal walk offers tranquillity but can also provide a sense of danger and bleakness'.

The resulting tracks on *A Pocket Guide to Dreamland* are also somewhat schizoid, veering between the melodic and rhythmic and the more desolate and abstract, capturing the unique fading atmosphere of the locations. This is, to borrow Lily Le Brun's description of photographer Martin Parr's work at the coast, about the *embodied* rather than the intellectual pleasures of the seaside.[8]

Tracks like 'Rollercoaster Ghost' begin with the slow, tense ascent of an old wooden scenic rollercoaster, chunking gears setting the rhythm, before we hear the screams of fear and excitement as its descends, while 'Sizzling Hot Dogs and Burnt Onions' summons the cacophonous atmosphere of a manic funfair, wailing sirens and electronic rhythms suggesting something tribal in the behaviour of holidaymakers imploring the ride to go faster. On 'Spectral Pier Ballroom' ska rhythms give way to the cavernous echoes of decayed big band music from days gone by, musically not unlike The Caretaker, all accompanied by the sound of relentless rain on the roof, a motif that leads into 'Penny Arcade in the Rain', on which Pink Floydesque sounds of money dropping into an arcade machine create a skittering tempo over which we hear trip hop samples.

Clarkson argues that his LP captures the multisensory nature of the coast, with 'colours and smell entering the creative process and affecting the mood of the music as they would in landscape art'. This reference to synaesthesia, and the relations of sound, vision and smell, is a repeated emphasis among many of those making music inspired by landscape. But whether these relationships are necessarily understood by listeners is another matter. For instance, our reactions to the woozy organ on the first track (Organ Donor), the sampled sounds of rollercoaster riders screaming, or the cawing of seagulls as they scavenge through discarded chip papers, depends on our own experiences of days out at the coast. For some the album may usher in a nostalgic rush of associations, memories of daytrips to the coast with families or sunburnt late-night stumbles through backstreets searching for a B&B. But for others, the album might seem all together too kitsch, a summoning of a grotesque working-class spectacle ultimately shunned by the middle-classes, a repressed memory that needs exorcizing. Desire and disgust are both in play, with the album working through questions of memory and loss that tap into class identities. Certainly, the association of coastal resorts with working-class rituals oftentimes justifies discourses of regeneration that suggest the only way to improve their prospects is to harness the civilizing influence of the 'creative' middle-class.[9] Notably, on Clarkson's album the opening track 'Organ Donor' returns as the last track 'Organ Transplant', but this time the organ wheeze and rollercoaster rattle bleed into the noise of construction work, with pneumatic drills and saws buzzing. Is this the sound of coastal gentrification, perhaps?

Such references to gentrification remind us that the coast is ever-changing (or 'liquid' to use Daniel Burdsey's particularly apposite term), with simplistic ideas of seaside revival based on arts-based regeneration obscuring more complex class conflicts.[10] In this way, Clarkson's album is akin to sociologist Bethan Prosser's auditory exploration of the gentrification of Brighton and Worthing, in which residents fixated on absent and lost sounds, such as the chatter of people on the beachfront or the noise of boats being winched ashore, noting these have been replaced by the incessant sound of construction: scaffolding being erected or the constant traffic noise associated with in-comers ferrying kids to and from out-of-catchment schools, morning, noon and night. As Prosser concludes, the 'gentry' that descended on Brighton and Worthing in the eighteenth century to transform them into fashionable resorts imbued them with a soundscape quite different from that of the contemporary gentrified seaside, with the sonification of the coast helping us think about the displacements accompanying regeneration.[11]

Such pressing questions of coastal gentrification are also broached by Heavy Cloud (aka Cornish-based Ryan Hooper), an act whose practice involves multimodal explorations of how the 'external' material landscape shapes the personal, 'internal' one. As a sound artist, Hooper mixes collage methods, poster production and textual exploration with sound poetry and audio recording, sometimes based on driftworks along the Cornish coast. As he tells me:

> The act of walking serves as a form of inspiration to me . . . the act of motion, matched with the act of looking, inspires me to soundtrack these journeys across actual and cerebral space. My music engages with my immediate local geography, and the wider county of Cornwall as well. It's not necessarily a conscious thought, but I sometimes feel I am an outsider encamped on a metaphorical island, making my little sound waves and sending them out to sea for a few people to find. And living in a county so often marginalised, and one equally proud of its independence, it's probably quite likely, however subconsciously, to have had my Kernow roots rub off on me. I wouldn't call my music necessarily rural, pastoral, or Celtic at heart, but . . . local geographies and histories have an influence over what I create.

The album *Atal Mor* (2022), for example – Cornish for 'flotsam and jetsam' – is essentially a four-day diary of walking along Cornish coastlines, rendered in droning synths, manipulated sounds and distorted voice memos, inspired by William Basinski's *Disintegration Tapes*, the films of Mark Jenkin, the writing of Wyl Menmuir and Philip Jeck's imaginary soundscapes. Notably, Hooper also credits Gwenno's *Tresor* album for reigniting his interest in exploring his Cornish roots.

Heavy Cloud's debut album was released on James A. McDairmid's Mailbox label in 2021, since when Hooper has produced a series of digital and limited-run cassette albums on labels including Brachliegen Tapes, Wormhole Records and Histamine Tapes. The physicality of the tape format clearly appeals to Hooper, with much of his music seeming to 'naturally fit the idea of an A side and a B side of the cassette, along with its hiss, hum and warble, and natural break as you flip the tape over'. This interest in materiality and affect clearly extends to his approach to landscape:

> I draw inspiration from the different textures found in Cornwall. In nature: rocks, waves, trees, clouds, fields, rivers, soil. The man-made: public transport, buildings, libraries, gardens, museums, cinemas, derelict or abandoned spaces. People, conversations, photographs, letters – different lines of communication . . . I enjoy incorporating characteristics that may be deemed hauntological into my music: musique concrete elements, degraded spoken word, vinyl manipulation/noise and tape hiss, mixing old and new technologies, abstract noise, and landscapes set in nowhere time.

On *Atal Mor*, masts and rigging rattle noisily, bells chime urgently and waves crash on harbour walls with increasing ferocity, tarpaulin ripping in the wind. *Season Shift* (2022) likewise explores Hooper's relationship with Cornwall across the year, starting in spring, tracking 'changing nature and weather, as well as changes in myself', he tells me. He suggests that sometimes at the Cornish coast 'the surge of nostalgia is so strong it sweeps up the present in a series of waves that roll up and down the beach'. Much of his work accordingly uses ambient synths and drones to connote the more ethereal elements of Cornish landscape and memory, but on *Scree* (2023), this is replaced by an altogether noisier palate of abrupt loops, feedback, harsh industrial percussion and distortion across two sides of a twenty-minute cassette. The release's two tracks – 'The sun has burnt these rocks' and 'The great continuous ebb and flow of the tide has crumbled the boulders' – borrow their titles from the unabridged journals of Sylvia Plath. These titles, together with the noisy instrumentation, allude to the erosion of Cornwall's cliffs by the sea and the fracturing of the whole into multiplicity. But this is not just about the destruction of land, but the erosion of Cornish identity, a story of cultural loss, the pulverizing forces of nature analogous to the waves of in-migration and capital accumulation transforming a once-productivist landscape into a consumerist fantasy of gentrified seaside leisure for Boden-wearing Londoners.[12]

But away from the noisy industrial drone of Scree, Heavy Cloud's (2022) *Beyond the Last Beach House* perhaps best encapsulates a hauntological approach to coastal gentrification (Figure 9.2). This apes The Caretaker's studious 'plunderphonics', old 45s and 78s jazz records played as if weathered by the sand on the beach, a project fusing the 'past and the present' which sit together 'not harmoniously, but in terms of communicating with one another'. This gentrification blues (and accompanying art print and collage album cover) is informed by a 'rush of inspirations' documented by Hooper: 'the painting of a row of beach houses I look at every day, visits to the South-West coast (past and present), beachcombing, the waltz and old ballroom songs, hauntology and The Caretaker, re-reading love letters, old film, young romance, vintage postcards, the golden age of cinema, fading dreams and memories of my grandmother'. As in Will Yates' Memotone project, focused on an old family holiday home in St Gabriel's, Dorset, here there is a sense here that the coastline itself is 'alive and moving, creating and disturbing – sometimes destroying – as it goes'. Social and

Figure 9.2 *Beyond the Last Beach House.* Courtesy Ryan Hooper.

physical change at the coast get equal billing: the release an effective distillation of the forces of change on the south-west coast, mirroring Marc Jenkin's (2019) film *Bait* in using obsolete, physical technologies and techniques to hint at things forgotten and buried.[13]

Coastal decay, national decline

Coastal gentrification is a major issue facing many coastal communities in a post-COVID-19 Britain where both housing and land remain in short supply, but the coast offers some of the most affordable housing. But on some parts of the coast, there is also a more existential threat: rapid sea level change and coastal erosion. In 2020, the US think tank Climate Central produced an interactive map showing that by 2050 up to a quarter of the English coast will be below the annual flood level, with projected sea level change of up to 30 cm meaning many coastal defences will no longer be fit for purpose. Even best-case scenarios predict this global warming will continue to at least 2100, suggesting major adaptations will be necessary, including the relocation of those living in areas of high risk. But on some parts of the coast, this is already happening, with dramatic erosion on the east coast a portent of things to come, roads, homes and caravan parks sliding into the sea as soft cliffs are undermined by winter storms. Ironically, much of the development in these areas was only made possible in the first place by coastal protection structures that aimed to 'fix' the land-sea interface.[14]

Many recent coastal driftworks – for example, Jules Pretty's *This Luminous Coast* (on East Anglia), Derek Turner's *Landfall: Edge of England in Lincolnshire* and my own *Borderland* (on Kent) – hence reflect on the fragile nature of the coast in the face of the predicted sea level change that threatens many towns and villages not in the distant future but within our own lifetimes. More widely, there is a major turn towards documenting the human and non-human geographies of coastal erosion through art and performance.[15] The soft erodable Norfolk cliffs around Happisburgh, for example, have attracted particular attention, currently being allowed to collapse by the authorities who have realized the battle against the tide is an expensive and losing one, with any attempt to defend the coast in one place potentially having devastating consequences for communities elsewhere given the dynamics of longshore drift and energy dissipation. In the winter of 2023–4 alone, more than ten metres of land was lost to the sea, with access to the village car park and beach destroyed. In this context, David Matless highlights the work of artists Julian Perry, Bettina Furnee and Malca Shotten, among others, as envisioning what might happen to caravans, bungalows and chalets as erosion continues, suggesting the emergence of a 'sea level melancholy' that's becoming fast-integrated into the modern histories of the Anthropocene.[16] This melancholy is also present in a track on Norfolk-based musician Luke Sanger's 2020 album *Global Horizontal Irradiance*, 'Happisburgh Annihilation'. This is a somewhat contemplative piece considering its title, an organically conceived modular synth track that emerges serenely from the confused signals of radio transmissions – pleas for help, perhaps? – before collapsing in a bewildering babble of bleeps. Elsewhere, Sanger (aka 'Luke's Anger') has produced work inspired by his walking on the wild stretches of the Norfolk coast, as well as the flatter inland areas of the fens, a landscape almost entirely shaped by successive attempts to keep the waters at bay: his album *Fool's Fayre* (2023) alludes to past times when these low-lying lands remained mysterious and miasmic, perhaps foretelling of a time when the unruly sea will reclaim the land again.

Up the coast, the low-lying areas of Holderness in the old East Riding of Yorkshire, between Spurn Head and Flamborough, endure what are thought to be the highest rates of coastal erosion in Europe. The 60 km coast consists of low cliffs of soft boulder clay, now retreating at a rate of around two metres per year. Since Roman times, it's thought as much as 5 kilometres of land has been lost to the sea, claiming an estimated thirty-two towns and villages including major communities such as Ravensar Odd. Established in the ninth century, this was once a bustling port and medieval new town, replete with tanneries, boatyards and marketplace. In the 1330s, severe storms damaged the quayside beyond repair, with residents moving to Hull or Grimsby in search of work, and the settlement all but abandoned fifty years later as the town's church was destroyed by the rising tide, graves looted and bodies washed out to sea.[17] Six centuries on, it still remains uncertain where this major community was located, but its memory lingers on in folk tales, a 'Yorkshire Atlantis' whose story contributes to the ongoing trauma of sea level change.[18] Today, other villages on the Holderness coast risk the same fate. For example, Skipsea, population 700, stands at risk of imminent destruction: many homes are boarded up and abandoned, their residents facing the indignity of being charged £6,000 for their demolition by the council as they leave. As Adam Farrer recounts in his prize-winning memoir *Cold Fish Soup*, 'the whole coast is defined by what's vanishing, and by what used to be there'.[19]

Holderness, as a 'country splitting into bits', is the subject of Conflux Coldwell's album *The Phantomatic Coast* (2021) and his accompanying film *Views from a Sunk Island* (2021). Conflux Coldwell (aka Michael Schofield, a Lecturer in Film, Photography & Media at Leeds University whose PhD was on 'The Hauntology of the Rephotographic Image') explicitly focuses on the ghostly properties of imagery and sound, and the way the passage of time can be registered technologically (Figure 9.3). For this 'amateur time traveller', the Holderness coast holds considerable attraction with its roads to nowhere, caravan parks teetering on the edge of cliffs, and old sea defences wrecked on the beach below. As Schofield explains 'The East coast of the country is land living on borrowed time, time we borrowed from the sea, reclaimed from marshland a thousand years ago . . . but now the sea has come to claim it all back'. His aim is to not simply document the erosion at the coast but explore the troubled relationship between land and sea, the ghosts of things lost, and a metaphysics of coastal change that gnaws away at the certainties of land and nation. The film is a montage of striking high contrast

Figure 9.3 Conflux Coldwell. Courtesy: M. Schofield.

black and white images, accompanied by the ghostly click of a slide projector and an experimental ambient soundtrack including field recordings made along the coast. As well as images of sandy shores and offshore wind turbines, we see beaches littered with concrete monoliths, the Second World War pill boxes and the other remains of military defence, a bunker archaeology waiting to be reclaimed by the sea. This emphasis on the landscapes of military defence at the coast suggests there is a selective memory in play which remembers somethings and forgets others, celebrating and memorializing the coastal defences that stood firm in times of war, but which are now sliding into the sea to join long-forgotten Saxon villages and ports.

This audio-visual project was partly inspired by Coldwell's purchase of Thomas Sheppard's (1912) *Lost Towns of the Yorkshire Coast* in a Bridlington junk shop for the princely sum of 55p. A map from this is reproduced on the gatefold album sleeve alongside multiple images of caravan parks and abandoned ruins, with particular emphasis on military structures that litter this crumbling coast: the wreckage of Godwin's Battery at Spurn Head, pillboxes at Skipsea and the now-inaccessible nuclear bunker at RAF Bempton, where eighteen-year-old Russell Behling went missing in 2010 amidst talk of satanic rituals. Following Coldwell's assertion that time and the sea destroy everything, the album crackles and hums with field recordings of waves crashing on beaches, fog signals and children playing on the sand sculpted into swooping, droning loops of sound, all overlaid with layers of static as if they are transmissions from years gone by. Track titles allude to ships lost to storms ('HMS Albion', 'Diana in the Ice') and the land that now lies beneath the North Sea – 'Doggerland' – once the Neolithic bridge between continental Europe and Britain.

Exploring themes of isolation, defence, heritage and nostalgia, Conflux Coldwell's sonification of the Holderness coast is a nuanced exploration of the fragility of the 'island nation' that arguably connects fears about coastal retreat to themes of post-imperial contraction. Focusing on those communities currently at risk, as well as those already lost to the sea, Coldwell makes uneasy connections between themes of military invasion – so central to the imagination of the English nation as 'fortress' – and the relative speed some parts of that nation are being lost to the forces of nature. These discourses resonate elsewhere on the East Coast. In Suffolk, for example, the vulnerability of the coast to aerial and naval attack in the 1940s led to the North Sea being depicted as the county's 'insatiable enemy': in her overview of the representation of the Suffolk coast in literature and art, Sophie Davies hence finds repeated reference to isolation and helplessness.[20] As

Mark Fisher himself also noted (in his K-Punk blog), Suffolk is a county that's important in national imaginaries because of its role in military history, and its beaches, like those of Holderness, are littered with unused defensive formations like pillboxes and Martello Towers. But Fisher also noted that Suffolk is fighting a losing war against the sea, with the gothic ghost stories of M. R. James being premonitions of the sea's ultimate victory over unstable land.

Among the many ghosts that haunt the Suffolk coast are those of the 300 or so who lost their lives in the North Sea floods of 1953, possibly the most devastating to affect England in 500 years. When experimental musician, field recordist and performer Matt Shenton moved to the Shotley Peninsula in Suffolk in 2015, the floods were one event he seized upon that allowed him to 'pull back the veil of the pink chocolate box cottages to expose the county's darker underbelly' (Figure 9.4). He tells me:

> The idea for the album *Stryance* came to me in February 2022 whilst reading a copy of *This Luminous Coast* by Jules Pretty. Pretty's book is a visionary look at the history of the East Anglian coastline, written after he walked its 400-mile path over a decade ago. In Chapter Two, he writes about the floods of 1953 where 'the sea walls were breached in over 1000 places, and 300 people died in the dark.' The flood hit during the dead of night. People woke up to find their possessions drifting and bumping about by their beds. Waves slicked silently across the land in some places; they exploded through windows in others. It would take days for newspaper reports of the disaster to reach the rest of the country.

Figure 9.4 there are no birds here. Ccourtesy: Matt Shenton.

The floods wreaked particular havoc where the sea surge burst through the banks of the River Orwell, tearing across Trimley Marshes and inundating Felixstowe's low-lying West End area. Forty-three were killed there, mainly prefab dwellers. Here Shenton makes connections between environmental and social injustice:

> There is a memorial to it in Felixstowe where there used to be prefabs: the most deprived areas were prefabs. They were supposed to last four years but 10–20 years on they were still there. It was the most vulnerable that suffered because they'd just been dumped on land that people must have known was going to flood because there were sea defences that had failed before the 1953 floods. Think about Jaywick, the most deprived area in Europe, and there's still people living in prefabs who will be some of the first affected by sea-level change. And should we be protecting them or should we just be letting the sea do its thing?

The music on *Stryance* mixes manipulated lo-fi field recordings with the sounds of modified electronics saved from landfill, with the resulting compositions textural explorations of timbre characterized by instinctive, chance-based arrangements:

> There's a lot of tape hiss on there mimicking the sea. But I wanted it to almost be like a bit of a panic attack: it kind of builds and I wanted to get that as when I was reading the book I kind of felt that my head was spinning and it was that sort of pressure. And so what you get is the pressure maybe of waves building against the window and then smashing through. I've never done a long form piece and I was quite interested in seeing whether I could do that. So it just worked like that, using Ableton to sort of make the hiss slowly build up, until three minutes from the end it's absolutely slamming into you, before it fades.

As he tells me, he did not want to dwell on the deaths that happened in 1953 or profit from their misfortune but wanted to follow Pretty's book in conveying how the lack of warning made so many unaware of the impending flood.

Shenton notes that Suffolk considered branding itself 'The Curious County' in 2012, deriving much of its strangeness from a coastal landscape of nuclear power stations, crashing container ports and villages falling into the sea. It's this general weirdness that encouraged Mark Fisher and sound-artist Justin Barton to create their audio essay (*On Vanishing Land*) based on a 2006 walk through coastal Suffolk from the port of Felixstowe to the inland Anglo-Saxon burial grounds at Sutton Hoo. On this album, washes of ambient sound specially composed by Gazelle Twin, Raime, Ekoplekz, Baron Mordant and others accompany Fisher's description of landscape features. Peeling back the veneer of the present, Fisher found a history of 'enigmatic departures': traces of the supernatural, myths from the time of the kings of East Anglia, and everywhere the 'ghosts of exteriority' which can be traced through the novels of M. R. James through to the more recent and sinister events at the Cold War installations of Orford Ness. When Sebald encountered the latter, at the remains of the Atomic Weapons Research Institute at Orford, where nuclear weapons were stress-tested until 1972, he wrote, 'the closer I came to these ruins, the more any notion of a mysterious isle of the dead receded, and the more I imagined myself amidst the remains of our own civilization after its extinction in some future catastrophe'.[21]

Sebald's melancholic travels on the Suffolk coast, and his uncanny sensation of slipping into a 'different time', have since become part of Orford Ness' strange history, forming another 'layer of significance' on top of existing tales of the monstrous such as the merman-like creature fished out of the sea in the late twelfth century, still spoken of as the spirit of a strange, unstable land.[22] In the 1980s, there were frequent UFO sightings over the local countryside which were thought to be associated with Ministry of Defence experiments with radar, ballistics and the atomic bomb but were probably caused by the glare of Orford lighthouse: irrespective, for a short time Orford acquired a reputation as Britain's 'Area 51'. There are also forgotten histories of indentured Chinese labour building the seawall around the military site in the early years of the twentieth century, alongside German prisoner of war labour during the First World War. A rumour also has it that the Nazis landed at the tip of Orford Ness in 1943 only to be driven back by an unexplained 'wall of fire' (a rumour scotched by official documents released fifty years' later).

A ten-mile spit of shifting shingle and mudflats whose lighthouse was demolished in 2020 because of the impending threat of coastal inundation, Orford Ness hence offers a dense and layered history of ghosts and memories that continues to attract hauntologists, bunkerologists and psychogeographers as much as ornithologists and dogwalkers. Since its purchase by The National Trust in 1993, multiple musicians and performers have converged on the Ness, 'each producing a different and distinctive mediation of what they see as its qualities of mystery, secrecy and intrigue'.[23] One of the first was Brian Eno, whose *On Land* (1982) references Orford Ness, while exteriors of various buildings on Orford were used in the video for 'Wires' by Athlete (2005), with lead singer Joel Pott walking among the empty, despoiled ruins of the military test site. But the number of albums about the area multiplied dramatically after Mark

Fisher's visitation. For example, on Drew Mulholland's (2020) *A Haunting Strip of Marshland*, a soundtrack to Robert Macfarlane and Adam Scovell's short film *Ness*, the Suffolk coast is sampled, rewound and remixed to Macfarlane's evident approval:

> Drew Mulholland knows how to listen to a place. How to lay ears to a landscape and catch the voices and utterances that surface into the skull. But fidelity is not his aim here – and nor should it be with such a deceiving, treacherous site as The Ness. The tracks you hear are – in their distortions and manipulations – truer to The Ness's unsettlements and resonances than any others I know. They hover somewhere between field-recording and haunting. Acoustic elements have been worked over until they are scarcely recognisable, surviving scantly as hyper-mutated catalysts. Mulholland even used the place itself as the distortive force; he gathered lichen species found on The Ness, ground them nearly into dust, glued the resulting powder onto a length of cassette tape, played and recorded the lichened tape, then stretched the recording to ten times its original length – before splicing it and adding reverb and echo. The resulting tracks are fabulously unfaithful to their origins . . . They bleep, shiver, rattle, quiver, shudder, detonate.[24]

The result is a soundscape evoking the Orford Ness landscape beneath and beyond the thresholds of visibility, hinting at its secret history as a site of atomic weapons testing. It also testifies to Mulholland's work as the 'godfather of hauntology', continuing a line of releases stretching from *Warminster* (1999, as Mount Vernon Arts Lab) through to *A Trip to Hob Moor* (2021).[25]

The Secrets of Orford Ness (2020) by Iain Chambers, member of music concrète collective Langham Research Centre, was also recorded among the old Ministry of Defence buildings, conveying a sense of its decay and the fact these are now overrun by nature. Air ducts that were once used ventilate missile testing laboratories are now nesting sites, and plants have taken root inside the facility. Subtitled a 'post-apocalyptic walk', the thirty-minute collage of field recordings begins abuzz with the sound of insects and birds but punctuated by the regular roar of US jets from RAF Mildenhall, and ultimately gives way to metallic clanging and ominous electronic chirps, as if we are listening to radio transmissions from the past. Throughout, a regular dripping tap marks the passing of time. Alongside other field recordings from Chris Watson and Brian D'Souza, this was played to visitors to Artangel's 2021 *Afterness* installation in the Black Beacon, the rotating wireless beacon constructed in 1929 to allow for more accurate marine navigation around Orford. The viewing apertures around the Beacon allowed visitors to survey the Ness landscape while listening to the recording, contemplating the connections between the past and futures of this previously hidden landscape.

The interlacing of the military-industrial and the natural-historical is also explored by Oliver Cherer (aka Gilroy Mere) and Robin Saville (from Isan) whose *Sizewell* (2019), combines field recording, folk guitar and electronics to evoke the nuclear uncanny some ten miles up the coast from Orford Ness. Saville lives nearby, and is acutely aware of the impact the proposed Sizewell C might have on RSPB Minsmere and the quiet town of Leisten:

> I have had a very conflicted relationship with Sizewell power station, but what used to really rankle with me was the fact that local artists quite often appear to be portraying the landscape with their backs to Sizewell. You know it's there in the landscape, there is no denying its existence. If you go out walking around here, it's quite hard to get lost because all around you are pylons and power lines all leading back to one place. And you know exactly where they're all leading to. So it seemed necessary to at least acknowledge it.

Cherer, on the other hand, based on the south coast in Hastings, knew Sizewell (Figure 9.5) mainly from family visits to the area, and vicariously through Brian Eno's soundscapes of the local area (confessing 'I've not read Sebald though everyone tells me I must'). On their multimedia collaboration, field recordings and photography guided the recording process, Cherer explaining:

> I allowed the field recordings to dictate the music . . . There were some less obviously appealing noises which I found increasingly compelling that I might ordinarily have edited out but were instead used as a spur to electronic intervention. Raucous chains dragging over machinery as some kind of industrial work was being done on the beach became an important part of one piece while the hum of the power station became part of an electronic drone, mixed with seagulls, kittiwakes, sea and wind.

The result is an ominous soundtrack, hinting at 'end of the line' weirdness. In their accompanying booklet, they write 'All the time there is that hum, low enough to allow it to be zoned out, but loud enough that, were it to stop suddenly, not only would we notice, but it would probably sound like the last split second before the end of the world'.

Construction on Sizewell C began in earnest in 2024, despite concerned residents pointing out rising sea levels and more regular storm surges would

Figure 9.5 Sizewell. Courtesy: Ted and Jen, CC BY-SA 2.0.

make the nuclear power station more susceptible to flooding than predicted, with the owners, EDF Power, only modelling coastal change up to 2140, even though spent nuclear fuel will remain on site for decades, perhaps centuries, longer. These concerns are very real ones given the histories of flood events on this coast, with the ancient capital of East Anglia, Dunwich, a small coastal village but once a sizable port town, just seven miles North of Leiston. In the eleventh century, it was the tenth largest town in England, hosting a large Franciscan priory and no less than eight churches. Its sheltered harbour was also the most important along the entire East Coast, but storm surges in the fourteenth century caused the River Dunwich to be diverted up the coast, and the great storm of 1387 swept 400 homes into the sea. The town was largely abandoned and is now something of a pilgrimage site for 'melancholy poets, medieval historians and researchers of coastal climatic change'.[26] 'Dunwich, with its towers and many thousand souls, has dissolved into water, sand and thin air', wrote Sebald, continuing, 'If you look out from the cliff-top across the sea towards where the town must once have been, you can sense the immense power of emptiness'.[27]

Dunwich is also the subject of *Gilden Gate* (2023) by Oliver Cherer's alter ego Gilroy Mere. Like his earlier projects as Gilroy Mere on Clay Pipe Music, this is a melancholic celebration of a vanished location, its story revealing itself to Cherer during the making of *Sizewell*, when he stayed at the Coastguard's Cottage at Dunwich. The release is split between a 'land-side' – breezy, freewheeling journeys along the sun-kissed coast and green lanes of Suffolk, tinged with pastoralism – and a 'sea-side' that heads beneath the waves, tracks named for the ruined priory and some of the churches now beneath the sea. In contrast to the pastoralism of the first side of the album, the second offers more contemplative explorations of marine memories, hesitant scratchy violin over growling drones and always the sound of crashing waves. Frances Castle's sleeve art accordingly accentuates the contrast between the land above and the imagined world beneath the sea, with church spires barely discernible beneath the inky blue waves. Like *Sizewell* before it, the project was guided by field recording:

> I recorded a lot of sea noise for *Gilden Gate*. Most of this was in situ. I did already have some useful field recordings from the *Sizewell* project . . . and used one or two of those but I didn't feel bound by them. There are also some church bells recorded locally in St Leonards [*Sussex*] which I muffled electronically to 'submerge;' them. There is always a certain tension between honouring the field recordings and intervening as an artist but I have to say that I'm more interested in an interesting piece of music rather than faithful representation. It's art after all.

Gilden Gate appears to tap into the current interest in sea level change, displacement and coastal abandonment (Figure 9.6), with Stefan Skrimshire writing, 'the story of the fall of Dunwich holds a unique kind of fascination, most likely for its resonance with our own contemporary relationship with the encroaching sea'.[28] Yet making such parallels was not necessarily Cherer's overt intent:

> The climate change analogy is an unavoidable one I think. It's not what drew me to the subject initially, but it occurred to me very quickly and, if I'm honest almost made me ditch the project. I didn't want to be heavy-handed about it in a way that quite a lot of political art tends to be. I felt that just putting it out there was enough and that making more of the climate change connotations would be painfully labouring the point.

But even if Cherer's work doesn't seek to make any specific political points about climate change, this album contributes to the range of representations that has established Dunwich, and the Suffolk coast more generally, as an iconic coastal landscape that makes the past available to us.

Figure 9.6 Dunwich. Courtesy: Steph Gray, CC BY-SA 2.0.

Living on the frontline

While the coast of East Anglia has become emblematic of anthropogenically induced climate change and the threat of sea level change, it's the southeastern corner of England that has featured most prominently in recent media representations of the fragility of the 'island nation'. Significantly, in the run-up and aftermath of the Brexit vote, the English Channel became depicted as the existential boundary between the nation and continental Europe, irrespective of the fact that the political border exists at any number of ports, airports and railway termini both within, and beyond, the shores of the UK.[29] But in the febrile context of discussions concerning the future of the UK out of Europe, the media's preoccupation with refugees landing in small boats on the beaches of Kent encouraged many to visit this coast, whether to offer a warm welcome to scared arrivals, or attempt to 'turn back the boats'.

In the post-Brexit era, when national identity has been subject to particularly intense scrutiny, concern about these landings – discoursed by Farage, Braverman and others on the right as a de facto 'invasion' by economic migrants – has made the Kent coast a potent site in debates around national values. This is not necessarily new given the towering chalk white cliffs of Kent have been symbols of national identity since at least the nineteenth century. Famously, Matthew Arnold's 1867 lyrical poem 'On Dover Beach' employed the cliffs as a metaphor for the national values he felt were under threat, invoking both patriotism and faith via reference to the 'calm cliffs' protecting from the 'grating roar' below. By the twentieth century, they were firmly woven into the national psyche as the gateway to England: writing in 1920, travel journalist Walter Jarrold memorably termed them 'the white walls of Albion' (England's original Roman name of Albion derived from the Latin *albus* for white). The First and Second World Wars consolidated this mythology: for soldiers on continental battlefields, the idea that they would once again view the white cliffs was pivotal in rituals of military departure and homecoming.

Given the persistence of these myths of military endeavour and national sacrifice, it's not surprising that patriotic representations of the white cliffs of Dover have been invoked by those arguing that severing ties with the EU was the best thing Britain has ever done. For many, they have become iconic of Britain's elemental insularity, propping up the idea of the 'island nation'. More than a mere icon of Englishness, here the cliffs reinforce a sense of *islandness*, and a distancing from continental Europe that's both physical and cultural. In this sense, Remainers were only too ready to seize upon a major cliff collapse in February 2021 as symbolic of post-Brexit decline, as if the very chalk itself were attempting to leave for Europe.

This type of metaphor resonates powerfully because of Kent's historical role as the military 'frontline' at many points in British history. Nowhere is Kent's defensive role more obvious than at Dover itself, where Britain's largest castle looms over the docks. Long described as 'the key to England', its immense walls enclose a Roman lighthouse, an impressive Norman keep and labyrinthine Napoleonic tunnels. Sold as part of the 'White Cliffs Experience', a series of multimedia exhibits describe how these tunnels were used for masterminding the 1940 Dunkirk evacuation when a flotilla of 800 boats rescued 300,000 British troops from the beaches of Northern France. Part of the Cold War Civil Defence Corps control room has now been turned into an 'escape room', challenging participants to secure the bunker before enemy missiles strike. Another part of the tunnels has been turned into an interactive exhibit where a nurse guides visitors through rooms documenting the journey of an English Second World War pilot fighting for his life after having been shot down by German planes.

In many ways this signifies the persistence of a 'wartime consciousness' at the Kent coast that links local senses of place to past military endeavours and national identities. Walking the Kent coast, war haunts every footstep, with the preservation of relict military landscapes – pillboxes, gun emplacements, batteries, Martello Towers, Cold War shelters – feeding into the sense that this has always been the frontline. One of the lesser-known, and weird, military remnants on this stretch of coastline are the large concrete sound mirrors which are found at several points along the same stretch of coast where refugees regularly land, including arrays at Dungeness, Hythe and Folkestone. This forerunner of radar resembles a 'listening ear', designed to make forms of distant noise audible (approaching aircraft engines, for instance) that would otherwise exist beyond the range of human hearing.

Once part of the London Air Defence warning system, sound mirrors remain an unsettling presence in the landscape. Like many of the other derelict military sites and ruins along this stretch of the coast, they have been enthused over by bunkerologists and brutalists alike and have become oddly iconic in recent years. For essayist and historian Brian Dillon, the fact the sound mirrors never functioned effectively because of ambient noise, being abandoned long before the Second World War, makes their meaning even more undecipherable. He writes of the mirrors as persistent reminders of the failure of Modernism, and the futility of trying to delineate the boundaries of the 'island nation' in a global era.[30] So when I ask the Kent-based act Degradation if they have any images of themselves and they send me one of themselves framed by the concrete parabola of the sound mirror at the top of Abbot's Cliff, between Dover and Folkestone (Figure 9.7), my interest is piqued.

Degradation is a collaboration between Dominic O'Donoghue (aka Vinegar Tom) and George Rayner-Law, the former living in Deal near the beaches of Walmer and Kingsdown, popular landing points for refugees just a few miles up the coast from Dover. Together, Dom and George run Brachliegen Tapes, an electronic/noise imprint specializing in cassette releases of experimental music with a strong industrial flavour. Some of their releases are inspired by anti-fascist bunkerology and the spectrality of landscape in general. Vinegar Tom, for example, released *The Pattern under the Plough* (2020) as an exploration of the buried memory of sound which lingers in images of remembrance. On 2021's *Do Not Lose Sight of the Sea*, Vinegar Tom also uses field recordings from the Kent coast (the dark edges) to weave two seven-minute meditations on coastal living, the nation equated to a behemoth sinking into the sea, 'sending out a distress signal to no-one, and blaming all those but itself'. The eponymous title track opens with 'rolling swells of shoreline static, and mocking seabirds, carried

Figure 9.7 Degradation at Abbot's Cliff, Kent. Courtesy: Degradation.

breathlessly on a hungry wind'. We hear electrical tones diffuse themselves, mingling with the tides of an uneven pulse, and, in the band's words, visit a coast where 'damp ghosts and hot piss linger in a fractured landscape, stretched and distorted by the pathos of national myth'.[31]

Living in Deal, Dom has been well-placed to comment on the drama unfolding in the channel, regularly seeing Border Patrol cutters searching for refugee boats, telling me that living in Deal 'you have that sense of being, if not on the border, then on the edge'. As Degradation, he recorded 2022's *Leadlined* with George, a response to the times shaped by the band's local environment but reflecting wider structures and relations, a musical collage of assembled thoughts and words about the state of the nation. Though frequently buried beneath a cacophony of manipulated field recordings, fragmented radio broadcasts, noisy overhead chinook helicopters and the sound of George's washing machine reverberating through his flat walls, snatches of vocals can be heard:

Possibility shrinks and shrinks
Degradation is all that is left
The fort needs defending
From the trash heap of time
Waste like dew against the sun
Rote response to media churn
The bridewell economy of Kentish men

Elsewhere, the lyrics allude to the topography and psychology of the Kent coast, a land of 'sand dunes, golf courses, beef dripping . . . country justice, pestilence, asbestos, industrial estates, police stations'.

The cover of *Leadlined* reflects these themes in an incredibly unambiguous manner. The image is by Simon Moreton, author of *Minor Leagues* zine and *Where* (2021), an acclaimed graphic novel concerning place, memory and the death of his father. The image is from one of his self-produced fanzines, the one-off pamphlet *What Is Britain?* (2020). This uses his trademark DIY style of photocopied images juxtaposed with fragments of maps, prose and poems which, in this case, present a polemic about 'the past inside the present; austerity, racism, colonialism, classism, poverty, middle-class apathy, and the cruelties of capitalism'. Like *The Lie of the Land* fanzine that preceded it, where Moreton remixes Watkin's ideas about ley lines, *What Is Britain?* explores the myth of Englishness, challenging the idea that the 'order of things is natural, and that those in positions of power have our best interests at heart'. In the zine, old images of thatched cottages and lazy rivers meandering through the countryside are juxtaposed with caricatures of David Cameron (with accompanying pig) and ghoulish UKIP councillors. But the cover image is particularly arresting, depicting one of Theresa May's anti-immigration 'Go Home' vans hurtling towards the edge of the cliffs of Dover, shortly to fall into the sea below (Figure 9.8). These Home Office vans were infamously sent out into London boroughs in 2013 adorned with the message 'In the UK illegally? Go home or face arrest'. Widely criticized as divisive, these nonetheless sent out the message that the Tory party was taking a tough line on illegal migration, stoking anti-refugee attitudes and partly laying the foundations for the decisive Brexit vote three years later.

Named *Leadlined* after the practice of lining coffins with metal to slow the composition of the body, the album seems a metaphor for the militarized bordering process designed to protect the rotting corpse of the nation. In the band's own words:

> Heaving drones of memory churn the decayed hum of modernism and the spiritual bus routes of the past. £40 million of military hardware flying low over the colonial jugular; the knifing wind off the Kentish coast, shearing through the tendons of Roman roads and Saxon shores. Through the record's mediation of these imagined pasts and displaced futures – the sinews of a settler class – the debilitating power of personal bereavement collides with the enervating deluge of a grieving national mythology. Through bricolage it attempts to come to terms with the inescapable, morbid nostalgia of the nation's psyche. In this impasse of postponed crisis, future horizons recede and misdirected resentments permeate all things.[32]

Figure 9.8 *Leadlined* by Degradation. Art by Simon Moreton, courtesy: Degradation.

Degradation are a low-key act who have received next to no media coverage. Their music is also low-tech, using a four-channel mixer plugged into a computer with some delay units, a microphone and an oscillator, all mixed with the ubiquitous Logic Pro. They are not trained musicians: tracks are emailed back and forth between London and the Kent coast, augmented and remixed. They tell me that there is little planning in this, no 'real intentionality' and that *Leadlined* was a process of 'piecing bits together that kind of came out as this this thing'. For them, the composition is in the editing. But they have an acutely sharp understanding of the place of landscape in the making of a national psyche. George tells me:

> The 'big dogs' of English Marxist historiography are very important in terms of how I see the world, that idea of Britain as a kind of imperial concept more than a place, and the idea that if the Britain of Imperial domination is in retreat or is ceasing to exist, what's left of the idea of England or Englishness? It's has been able to exist in a superposition for hundreds of years and then has had to move out of that superposition: what does that look like? Especially in the context of a country that's getting poorer. If everything is in retreat at the same time, what's it like to experience that?

Our freewheeling conversation in a seaside pub is marked with references to post-colonial melancholy and the work of E. P. Thompson, Paul Gilroy and Stuart Hall as much as to Nurse with Wound, Throbbing Gristle and Nocturnal Emissions.

Kent is no hotbed of experimental music or hauntology. But several key events in the history of the nation have happened here, which gives Degradation's *Leadlined* its clout. The Roman invasion began at Deal in 55 BC, while St Augustine landed in AD 597 at Pegwell Bay just ten miles to the North, reputedly bringing Christianity to Britain. In between, the arrival of the Germanic leaders Hengist and Horsa occurred in AD 449 on Sandwich Bay, the location now a golf course. Some consider Horsa and Hengist the first Englishmen, Anglo-Saxons who fought for the English king Vortigern against the Picts. George describes this to me as an important foundation myth in the making of nationhood:

> If Saxon-ness which the English state develops out of starts here and then expands to Britishness as we know it, then that's when an idea of an English identity is developed. It's an imperial identity. And then, if it's retreating, retreating, retreating, surely the last place it will retreat back to is the place it came from: the East Coast of Kent.

So if England is constituted by myths and legends, many of which still circulate in the present, it is at the coast we readily find these myths memorialized, albeit in a landscape fast eroding in the face of a realization the nation is not, and cannot be, what it once was.

Further Reading

Bolton, Tom (2018). *Low Country Brexit on the Essex Coast*. Penned in the Margins.

Davis, Sophie (2020). *Island Thinking: Suffolk stories of landscape, militarisation and identity*. Springer.

DeSilvey, Caitlin (2014). 'Palliative curation: Art and entropy on Orford Ness'. In Olsen, Bjornar & Pétursdóttir, Pora (eds), *Ruin memories*. Routledge, 79–91.

Fryers, Mark (2021). 'The haunted seas of British television: Nation, environment and horror'. *Gothic Nature*, 2, 131–155.

Hubbard, Phil (2022). *Borderland: Identity and belonging at the edge of England*. Manchester University Press.

LeBron, Lucy (2022). *Looking to Sea: Britain through the eyes of its artists*. Sceptre.

Matless, David (2017). 'The anthroposcenic'. *Transactions of the Institute of British Geographers*, 42(3), 363–376.
Morris, Benjamin (2014). 'In defence of oblivion: The case of Dunwich, Suffolk'. *International Journal of Heritage Studies*, 20(2), 196–216.
Pretty, Jules (2014). *This Luminous Coast: Walking England's eastern edge*. Cornell University Press.
Rees, Gareth (2024). *Sunken lLnds: A journey through flooded kingdoms and lost worlds.* Elliot and Thompson.
Turner, Derek (2022). *Edge of England: Landfall in Lincolnshire*. Hurst Publishers.
Wollett, Lisa (2024). *Lost to the Sea: A journey round the edges of Britain and Ireland.* John Murray.

10

Conclusion

In the early decades of the twenty-first century, as the economy stagnates and the nation fractures on lines of ethnicity, class and age, the vexed question of national identity remains at the forefront of many peoples' minds. What, precisely, does it mean to be English in the post-Brexit era? How does this relate to the identity of the wider British Isles and UK? And what are the imagined futures of England in a fast-changing world where national boundaries appear increasingly irrelevant to the resolution of an environmental crisis that knows no limits? These questions are ones that this book has addressed, albeit sometimes obliquely, through a consideration of the way the landscapes of England have been represented in contemporary electronic music, especially the other-worldly hauntological sounds associated with acts such as those on Ghost Box Records, as well as fellow disciples of analogue atmospherics and wonky electronica on labels including Clay Pipe Music, Castles in Space, Wayside & Woodland, Fenny Compton, Plenty Wenlock, among numerous others.

This book has not been overly fastidious in its definition of what is or isn't hauntology, a term some of those interviewed for this book distanced themselves from because of concerns about it pigeonholing their work in some way. But what is clear is that this genre – or micro-genre – is a distinctively British one, grounded in the cultural politics of the nation and, in the examples considered in this book, its contested landscapes. This book has hence traced the way that select musicians, sonic-artists and field recordists have engaged with these landscapes in the early decades of the twenty-first century, alighting on the idea that their music is particularly interested in the interplay of absence and presence: that is to say, the music is about the presence of things from the past that should no longer be there as well as the things that should be there but aren't. Through sonic manipulation, field samples, and the engineering of affect, hauntronica helps us see – or rather hear – the spectral presences that simmer just beneath the surface of the landscape.[1]

In this book, we have then heard of thwarted dreams of future cities, forged in the white heat of 1960s technology and town planning, as well as the forms of ritual and festival that cemented the links between people and place in the premodern rural communities that moved to the seasonal rhythms of nature. We have also heard the echoes of tribes gathering at stone circles and travelled with them on pagan pathways as we have marvelled at the strange numinous beauty of the ancient landscape. Reaching back into even deeper time, we have been encouraged to consider how our current economy, society and culture has been built on the bedrock of geology. Peeling back the layers of time, and oscillating between the weird and eerie, hauntronic music makes the familiar landscape strange indeed.

So does the upsurge of hauntronic music since 2010, and particularly since the Brexit vote in 2016, acknowledge a profound loss of faith in our ability to envisage a progressive future for the English nation? Is the retreat into the soundscapes and landscapes of the past part of a leftist melancholia that seemingly offers no route out of capitalist realism and the current political preoccupation with neoliberalism? The premise of this book has been that although the music that engages with the landscape through the lens of hauntology remains something of a niche concern, it represents a productive and oddly timely set of cultural resources for exploring the politics of place and nationhood. Far from embracing a 'restorative nostalgia' which simply aims to reconstruct the past in the present, this book has suggested that hauntronica is *radically* nostalgic, asking us to ensure the 'shattered fragments' of landscape and memory remain just that: fragments that do not necessarily cohere but which must co-exist.[2]

In effect, and perhaps optimistically, the book's argument has been that this sonic dis/orientation of the English landscape via forms of 'uneasy listening' is a form of cultural critique helping us forge a new radical national politics of inclusion, pointing the way to postcapitalist futures and less alienated forms of life. In this sense, this book has not just been a primer on experimental landscape music but has used this to highlight possible paths that might lead the nation away from post-Brexit national malaise, environmental catastrophe and economic precarity. So while most of the music discussed in this book could be dismissed as the outcome of a very personal and idiosyncratic engagement with the local environment, and sometimes arcane interests in architecture, archaeology, ecology or geology, collectively it speaks to the major issues of our time through a sonic invocation of the eerie, the strange and the displaced.

In a broader sense, this book has been a response to the often-voiced critique that hauntology and related practices of psychogeography and place-writing are, at best, apolitical and, at worst, a retreat into nostalgia. As Justin P. Hopper suggests, hauntological excursions into the landscape in media, music and writing must challenge dominant narratives and not enshrine any one individual's experience or insights as somehow 'correct'. In his own work as author of *The Old Weird Albion*, for example, and as part-time member of Belbury Poly, Hopper layers the landscape with different beliefs, 'following the paths of renegades' to depict places of heterodox belief, pagan survival and 'vanished prehistories of worship'.[3] This is no simple reclaiming of Albion based on the idealization of any one period in its history, but an attempt to make sense of the multiple experiences of landscape through a complex enfolding of space and time that is *heard* as well as seen.

One major caveat must be noted here. While there are some important exceptions – Hannah Peel, Rebecca Denniff, Frances Castle and Cate Brooks, to mention a few – the music profiled in this book is predominantly the work of middle-aged men. Many of the acts repeatedly referred to in this book – Belbury Poly, Craven Faults, epic45, Gilroy Mere, Spaceship, Vic Mars, Warrington-Runcorn New Town Development Plan and The Heartwood Institute – are, broadly speaking, all men of a certain age, part of the haunted generation that grew up in the 1970s. They now have time and resources which they can invest in the production of music that, for the most part, will never generate a significant income for them but which circulates through networks of similarly aged cognoscenti. This is not to say that hauntronica performs the type of masculine virtuosity or phallic power associated with, for example, heavy metal music, but this does raise questions about silences and gaps in the putative hauntronic scene. Importantly, where are the non-white performers using hauntological sonics to speak of the production of landscape as reliant on histories of slavery and Empire? Further, and relatedly, given hauntology frequently replays and reworks British popular musical cultures of the 1960s and 1970s, is it destined to ignore the strands of hip-hop, reggae, dub and Afrofuturism that were so often excluded from that media space? And why does so much hauntological music emanate from largely white creative enclaves like Todmorden, Lewes or Whitby? Where must we look to find black feminist hauntology?

These are all important questions, not least in the context of my argument that this music is no mere exercise in nostalgia, but a provocation that helps us re-imagine England's place in the world, the political ecology of the

Anthropocene and vital questions of multiculturalism. However, for now, I must set these aside and instead conclude by offering some schematic thoughts on the sometimes-overwhelming body of electronic music that has explored landscape via hauntological techniques and methods over the last decade or so. Here, I offer some brief final reflections on why this weird and wonderful landscape music must be considered more than the sum of its parts, symptomatic of a more general search for meaning in a world that sometimes appears to have lost its bearings.

The first is that the music discussed in this book is, to a greater or lesser degree, both experimental *and* enjoyable. Emphasizing its theoretical and even cerebral dimensions should not distract from this fact: even though a frequent trope is the elicitation of loss and the creation of an accompanying atmosphere of melancholia, the music has a therapeutic component, helping us re-orientate ourselves in the world, encouraging us to take time to listen to and explore our surroundings. It takes us away from the capitalist present, and the alienated conditions of our current existence, by engaging both our bodies and minds.

The second is that hauntronica does not just dwell on trauma and displacement but seeks to re-enchant the landscape. By foregrounding that which is lost in the present, but still haunts, it insinuates a politics of resistance that moves beyond questions of depression and grief to propose a different 'ethics of being', challenging the neoliberal present. Music that moves to the rhythms of the natural or 'green' calendar, for example, is part of this, an encouragement for us to consider the needs and wants of our bodies rather than accept the routines of everyday life, centred as they are on the imperatives of advanced capitalism.

Third, and finally, hauntological excursions in the landscape engage in an almost archaeological process of stripping back the layers of conventional representations of Englishness – the rural idyll, for example – to explore the social, cultural and political detritus that lies beneath, fragments often too weird or arcane to register as part of 'English heritage'. In so doing, hauntological music disrupts capitalist hegemony, identifying the 'cracks' in the landscape of Englishness. It raises the possibility that we might forge fairer, more inclusive futures than those embodied in the 'island nation' mythology peddled by those on the political right. It challenges their exclusionary nationalism by hinting at a commons that might be shared, and managed, by all.

These are perhaps bold claims for music that exists largely in the margins away from the Top 40, circulating in hyper limited-edition runs on tape and vinyl, certainly of little interest to radio broadcasters beyond a few specialist internet radio stations and, occasionally, a token play on BBC Radio Six's *Freak Zone*. But my conclusion that hauntronica represents a timely, critical engagement with landscape is one made on the basis that this is part of a wider 'hauntological turn' in the arts, literature and music that's beginning to infiltrate the cultural mainstream in subtle but meaningful ways. Slowly but surely, more and more people are embracing the weird and the arcane, tapping into folklore customs and creating meaningful connections with their surroundings.[4] They are Weird Walking and attending Stone Club; they are enthusing over the lost architectures of Modernity and buying into Brutalism; some are losing themselves in deep time, others lingering in the liminal spaces of subtopia or paddling in the rising tides of the English coast.

All of this suggests that hauntronic excursions into the landscape are part of a wider movement that is encouraging people to question the national heritage they are told to consume. Whether or not this hauntronica is more than a short-lived musical fad remains to be seen. But if, as Mark Fisher argued, 'we are induced by ubiquitous PR into falsely overestimating the present, and those who can't remember the past are condemned to have it resold to them forever', it feels like an overdue disruption of capitalist hegemony. In this sense, we should perhaps leave the last word to Fisher, who dismissed the idea that hauntology is a simple yearning for a vanished past by highlighting the complex stratigraphy of needs and wants irradiating hauntronic music. Commenting on the oneiric pop of Ghost Box, for example, Fisher argued that its citation of British horror films, Modernist architecture and pre-1979 television was not mere mimicry but an attempt to estrange capitalist realism, suggesting that the perceived 'gains' of neoliberalism should not be viewed as contingent on what was lost when the welfare state was effectively binned in the 1980s:

> We shouldn't have to choose between, say, the internet and social security. One way of thinking about hauntology is that its lost futures do not force such false choices; instead, what haunts is the spectre of a world in which all the marvels of communicative technology could be combined with a sense of solidarity much stronger than anything social democracy could muster.[5]

As he concludes, what is being longed for in hauntology is not a particular historical period, but 'the resumption of the processes of democratisation and pluralism'. if we listen carefully to the landscape, we can still hear the echoes of these processes, subsumed in the present. Dare we dream of an alternative Albion where they are allowed to flourish again?

Notes

Chapter 1

1 On the geographies of Brexit vote, see Morgan, John (2017). 'A cultural geography of Brexit'. *Geography*, 102(3), 153–160; Manley, David, Jones, Kelvyn, & Johnston, Ron (2017). 'The geography of Brexit–What geography? Modelling and predicting the outcome across 380 local authorities'. *Local Economy*, 32(3), 183–203.

2 On the politics of 'levelling up', see Jennings, Will, Lawrence McKay, & Stoker, Gerry (2021). 'The politics of levelling up'. *The Political Quarterly*, 92(2), 302–311.

3 King's College Policy Institute conducted research during the 2019 General Election campaign, finding people's Brexit identities as Leave or Remain voters had grown stronger since the 2016 vote, creating mistrust of those with alternative views. See https://www.kcl.ac.uk/policy-institute/research-analysis/divided-britain

4 One poll suggested 70 per cent of those describing themselves as English rather than British voted Leave. See https://blogs.lse.ac.uk/politicsandpolicy/the-black-box-of-brexit-identification-with-englishness-is-the-best-clue/

5 On Braverman's 2022 speech on asylum seekers at the coast, see my commentary at https://www.kcl.ac.uk/suella-bravermans-talk-of-a-refugee-invasion-is-a-dangerous-political-gambit-gone-wrong. In Hubbard, Philip (2022). *Borderland: Identity and belonging at the edge of England*. Manchester University Press, I detail why the metaphor of invasion was so powerful in Brexit times..

6 https://www.express.co.uk/news/politics/675034/chris-grayling-green-belt-migrants-countryside-eu-referendum-brexit-david-cameron-uk

7 Davies, Jonathan (2021). 'Brexit and Invasive Species: A case study of the cognitive and affective encoding of "Abject Nature" in contemporary nationalist ideology'. *Cultural Studies*, 36(4), 568–597. See also Berberich, Christine (2006). 'This green and pleasant land: Cultural constructions of Englishness'. In Burden, Robert & Kohl, Stephan (eds), *Landscape and Englishness*. Rodolphi, 207–224.

8 Garrard, Greg (2020). 'Brexit ecocriticism'. *Green Letters*, 24(2), 110–124.

9 See Loh, Lucienne (2013). *The postcolonial country in contemporary literature*. Springer; Fowler, Corinne (2021). *Green unpleasant land: Creative responses to rural England's colonial connections*. Peepal Tree Press.

10 Sethi, Anit (2021). *I belong here: A journey along the backbone of Britain*. Bloomsbury Nature.

11 Masud, Noreen (2023). *A flat place*. Penguin.

12 Harvey, James (2023). *John Akomfrah*. Bloomsbury.
13 Tolia-Kelly, Divya P. (2011). 'Narrating the postcolonial landscape: Archaeologies of race at Hadrian's Wall'. *Transactions of the Institute of British Geographers*, 36(1), 71–88.
14 Gandy, Matthew (2016). 'Unintentional landscapes'. *Landscape Research*, 41(4), 433–440.
15 Alex Nevin argues this selective imagining of Englishness has become more, rather than less, prominent in the twenty-first century, a 'cultural daydream of the neoliberal classes' invoked in a variety of reactionary cultural and political projects. See Niven, Alex (2019). *New model Island: How to build a radical culture beyond the idea of England*. Watkins.
16 Taylor, Peter J. (1991). 'The English and their Englishness: "A curiously mysterious, elusive and little understood people"'. *Scottish Geographical Magazine*, 107(3), 146–161.
17 Pitcher, Ben (2016). 'Belonging to a different landscape: Repurposing nationalist affects'. *Sociological Research Online*, 21(1), 77–89.
18 Matless, David (2010). 'Describing landscape: Regional sites'. *Performance Research*, 15(4), 72–82.
19 Gallagher, Michael, Kanngieser, Anja, & Prior, Jonathan (2017). 'Listening geographies: Landscape, affect and geotechnologies'. *Progress in Human Geography*, 41(5), 618–637.
20 Gallagher, Michael (2015). 'Sounding ruins: Reflections on the production of an "audio drift"'. *cultural geographies*, 22(3), 467–485.
21 Gallagher, Michael (2015). 'Landscape audio in situ'. *Contemporary Music Review*, 34(4), 316–326.
22 Revill, George (2018). 'Landscape, music and sonic environments'. In Howard, Peter, Thompson, Ian, Waterton, Emma, & Atha, Mick (eds), *The Routledge companion to landscape studies*. Routledge, 264–273.
23 For a summary of work on sonic affects in everyday listening, see Duffy, Michelle & Waitt, Gordon (2013). 'Home sounds: Experiential practices and performativities of hearing and listening'. *Social & Cultural Geography*, 14(4), 466–481.
24 Stahl, Geoff & Percival, Mark (eds) (2022). *The Bloomsbury handbook of popular music, space and place*. Bloomsbury.
25 See Saylor, Eric (2017). *English pastoral music: From Arcadia to Utopia, 1900–1955*. University of Illinois Press.
26 King, Richard (2019). *The Lark Ascending: The music of the British landscape*. Faber & Faber. See also Revill, George (1991). 'The Lark Ascending: Monument to a radical pastoral'. *Landscape Research*, 16(2), 25–30; Ross, Ryan (2024). 'Letting

the Lark Ascend: Vaughan Williams's "most popular work" and the limits of revisionism'. *Nineteenth-Century Music Review,* 1–24.

27 For a critique of this nostalgic English ruralism see: Mandler, Peter (1997). 'Against "Englishness": English culture and the limits to rural nostalgia, 1850–1940'. *Transactions of the Royal Historical Society,* 7, 155–175.

28 Revill, George (2000). 'English pastoral'. In Naylor, Simon, Ryan, James, Cook, Ian, & Crouch, David (eds), *Cultural turns/geographical turns: Perspectives on cultural geography.* Routledge, 140–158.

29 Quoted in Smiles, Samuel (1986). 'Samuel Palmer and the pastoral inheritance'. *Landscape Research,* 11(3), 11–15.

30 This phrase was possibly first coined by Lutyens in a Dartington College lecture in the 1950s, but its source is unclear.

31 Leidner, Felix (2021). 'P.J. Harvey as a modern war poet: How let England shake challenges "English England" through the pastoral'. In Gurke, Thomas and Winnett, Susan (eds), *Words, music, and the popular: Global perspectives on intermedial relations.* Springer, 111–132.

32 Gardner, Abigail S. (2017). 'P.J. Harvey and remembering England'. In Brooks, Les, Donnelly, Mark, & Mills, Richard (eds), *Mad dogs and Englishness: Popular music and English identities.* Bloomsbury, 66-82.

33 See also Matless, David (2023). *About England.* Reakton.

34 https://medium.com/the-riff/album-review-i-inside-the-old-year-dying-by-pj-harvey-cee9834a2a4

35 https://www.thewire.co.uk/in-writing/interviews/gazelle-twin-shares-new-album-and-interview

36 Burns, Will (2021). *The paper lantern.* Hachette, 67.

37 Ibid., 130.

38 https://www.manafonistas.de/2019/03/20/141005/

39 On rural gentrification see Phillips, Martin, Smith, Darren, Brooking, Helen, & Duer, Mara (2021). 'The agencies of landscape in rural gentrification: Impressions from the wood, the village and the moortop'. *Sociologia Ruralis,* 61(4), 778–807.

40 Ibid., 89.

41 https://joebedford.co.uk/will-burns/

42 Sleeve notes, *Chalk Hill Blue.*

43 Young, Rob (2011). *Electric Eden: Unearthing Britain's visionary music.* Macmillan.

44 See Luckhurst, Roger (2017). 'The weird: A dis/orientation'. *Textual Practice,* 31(6), 1041–1061.

45 Wheatley, Helen (2011). 'Beautiful images in spectacular clarity: Spectacular television, landscape programming and the question of (tele)visual pleasure'. *Screen,* 52(2), 233–248; Woods, Faye (2019). 'Wainwright's West Yorkshire: Affect and

landscape in the television drama of Sally Wainwright'. *Journal of British Cinema and Television*, 16(3), 346–366; Flanagan, Kevin (2016). '"Green and pleasant land"?: Ben Wheatley's British cinema between romanticism and modernism'. *Critical Quarterly*, 58(1), 16–22.

46 On the idea of the weird/wyrd in music, and particularly the relationship between (neo)folk and traditional folk, see Haddield, James (2022). 'Exploring the re-emergence of "weird folk"'. *Ele-king Magazine*, translated from Japanese and republished at https://tradfolk.co/music/writing-music/weird-folk/

47 See, among others, Cosgrove, Denis (1985). 'Prospect, perspective and the evolution of the landscape idea'. *Transactions of the Institute of British Geographers*, 10(1), 45–62; Mitchell, Don (2001). 'The lure of the local: Landscape studies at the end of a troubled century'. *Progress in Human Geography*, 25(2), 269–281; Daniels, Stephen (2011). 'What landscape means to me'. *Landscapes*, 12(2), 84–94.

48 Coverley, Merlin (2020). *Hauntology: Ghosts of futures past*. Oldcastle.

Chapter 2

1 See Scovell, Adam, 'Perception of a landscape in a Journey to Avebury', https://celluloidwickerman.com/2014/12/29/perception-of-landscape-in-a-journey-to-avebury-1971-derek-jarman/

2 https://www.cyclicdefrost.com/2009/04/belbury-poly-interview-by-emmy-hennings/

3 Interview with Jim Jupp by Mark Fisher, https://www.factmag.com/2009/02/04/interview-belbury-poly/

4 https://www.cyclicdefrost.com/2009/04/belbury-poly-interview-by-emmy-hennings/

5 Interview with Jim Jupp by Mark Fisher, https://www.factmag.com/2009/02/04/interview-belbury-poly/

6 https://www.theguardian.com/music/2009/jan/30/ghost-box-bellbury-poly

7 Interview with Jim Jupp by Mark Fisher, https://www.factmag.com/2009/02/04/interview-belbury-poly/

8 This unravelling of ontology through reference to the spectral is discussed at length in Burns, William (2025). *Ghost of an idea*. Headspace, wherein he argues 'ghosts, and spectres are metaphors representing individual and collective guilt and responsibility in the late twentieth century, a shared feeling of fear, awe, sorrow, and anticipation when faced with millenarian hopes and forebodings for the future' (16).

9 Miller, Jacob (2023). *Retail Ruins: The ghosts of post-industrial spectacle*. Policy Press.

10 Lown, Josh (2025). 'Spectres of gentrification: Towards a hauntological framework for exploring the impacts of gentrification'. *Urban Studies*, https://doi.org/10.1177/00420980241306O8

11 Leach, Rebecca (2024). 'Turning over stoke-on-trent (pottery): Ritual, affect and haunting'. *The Sociological Review*, 72(1), 76–98.

12 See, for example, Rahimi, Sadeq (2021). *A hauntology for everyday life*. Springer; Roberts, Elisabeth (2013). 'Geography and the visual image: A hauntological approach'. *Progress in Human Geography*, 37(3), 386–402; Shaw, Kate (2018). *Hauntology: The presence of the past in twenty-first century English literature*. Springer.

13 See Fisher, Mark (2012). 'Why mental health is a political issue'. *The Guardian*, July 16, https://www.theguardian.com/commentisfree/2012/jul/16/mental-health-political-issue

14 This essay is in Fisher, Mark (2014). *Ghosts of my Life*. Zero Books.

15 Interview with Leyland Kirby by Fisher, Mark (2009). *The Wire*, issue 304, June.

16 Fisher, *Ghosts of my Life*, 74.

17 Cited in https://thequietus.com/articles/04153-the-advisory-circle-mind-how-you-go-ghost-box

18 Fisher, Mark (2012). 'What is hauntology?' *Film Quarterly*, 66(1), 16–24.

19 Scovell, Adam (2017). *Folk Horror: Hours dreadful and things strange*. Liverpool University Press.

20 Fisher, Mark (2016). *The Weird and the Eerie*. Repeater.

21 Christodoulou, Chris (2018). 'Haunted science: The BBC Radiophonic Workshop and the lost futures of hauntological music'. *Scene*, 6(2), 107–120.

22 Fisher, *The Weird and the Eerie*.

23 Sexton, Jamie (2023). 'Producing paranormal sounds: Electronic music, projection, and blurred boundaries in *The Legend of Hell House* (1973) and *The Stone Tape* (1972)'. In Donnelly, Kevin & Mollaghan, Aimee (eds), *Haunted Soundtracks: Audiovisual cultures of memory, landscape, and sound*. Bloomsbury.

24 HRH Prince Charles (1989). *A Vision of Britain: A personal view of architecture*. Doubleday.

25 Hubbard, Phil (1995). 'Urban design and local economic development: A case study in Birmingham'. *Cities*, 12(4), 243–251.

26 Cited in interview with Broadcast by Joseph Stannard, 2009, reprinted in full at https://www.thewire.co.uk/in-writing/interviews/unedited-broadcast

27 Ibid.

28 https://thequietus.com/articles/09757-owen-hatherley-a-new-kind-of-bleak

29 Sutherland, Callum (2023). 'Mark Fisher and reimagining postcapitalist geographies'. *Dialogues in Human Geography*, 13(1), 99–118.

30 Mould, Oli (2023). 'Situating Mark Fisher's spatiality?' *Dialogues in Human Geography*, 13(1), 129–132.
31 Maddern, Jo & Adey, Peter (2008). 'Spectro-geographies'. *cultural geographies*, 15(3), 291–295.
32 This term is taken from Harvey, David (2007). 'Neoliberalism as creative destruction'. *The Annals of the American Academy of Political and Social Science*, 61(1), 22–44.
33 See for example: Sexton, J. (2012). 'Weird Britain in exile: Ghost box, hauntology, and alternative heritage'. *Popular Music and Society*, 35(4), 561–584.
34 https://www.cyclicdefrost.com/2009/04/belbury-poly-interview-by-emmy-hennings/
35 https://www.clashmusic.com/features/label-profile-ghost-box/
36 Pattie, David (2016). 'Stone tapes: Ghost box, nostalgia, and postwar Britain'. In Whiteley, Sheila & Rambarran, Shara (eds), *The Oxford handbook of music and virtuality.* online edn. Oxford Academic, https://doi.org/10.1093/oxfordhb/9780199321285.013.24
37 Thurgill, James (2020). 'A fear of the folk: On topophobia and the horror of rural landscapes'. *Revenant: Critical and Creative Studies of the Supernatural*, 5, 33–56.
38 Roy, Elodie (2014). 'Displacing the past. Mediated nostalgia and recorded sound'. *Volume!. La revue des musiques populaires*, 11(1), 145–158.
39 Gray, Chantelle (2022). 'A hauntology of clandestine transmissions: Spectres of gender and race in electronic music'. *Indian Journal of Gender Studies*, 29(3), 319–334.
40 Scovell, *Folk Horror: Hours dreadful and things strange*, 96.

Chapter 3

1 Preece, Ian (2020). *Listening to the Wind:encounters with 21st century independent labels.* Omnibus Press.
2 From Uncanny Landscapes podcast with Frances Castle (2023), https://uncannylandscapes.podbean.com/e/uncanny-landscapes-17-frances-castle-of-clay-pipe-music/
3 https://hauntedgeneration.co.uk/2021/12/08/frances-castle-stagdale-and-clay-pipe-music/
4 Watts, David (2004). 'Evaluating British railway poster advertising: The London & North Eastern Railway between the wars'. *The Journal of Transport History*, 25(2), 23–56; Savage, Rebecca (2020). 'The artist, the advertiser, the public: The great

Western railway poster as collaborative design'. *Visual Culture in Britain*, 21(3), 363–384.

5 Wright, Patrick (1995). *The Village that Died for England.* Cape.

6 https://hauntedgeneration.co.uk/2023/06/21/tyneham-house-and-clay-pipe-music/

7 Coates, Ken & Silburn, Richard (1970). *Poverty: The forgotten Englishman.* Penguin.

8 Mckenzie, Lisa (2012). 'A narrative from the inside, studying St Anns in Nottingham: Belonging, continuity and change'. *The Sociological Review*, 60(3), 457–475.

9 https://concreteislands.com/clay-pipe-music-from-the-foreshore/

10 https://www.caughtbytheriver.net/2017/09/green-belt-public-transport-music-oliver-cherer/

11 https://concreteislands.com/the-trains-have-gone-gilroy-mere-interview/

12 Matless, David (1997). 'Moral geographies of English landscape'. *Landscape Research,* 22(2), 141–155. See also Jeremiah, David (2010). 'Motoring and the British countryside. *Rural History*, 21(2), 233–250.

13 https://hauntedgeneration.co.uk/2020/02/29/gilroy-mere-oliver-cherer-and-over-the-tracks/

14 Love, S. (2021). 'The magic of the downs: Englishness in crisis in Eric Ravilious's chalk figure paintings'. *Aspectus*, 3, 1–18.

15 Harris, Alexandra (2010). *Romantic Moderns: English writers, artists and the imagination from Virgina Woolf to John Piper.* Thames & Hudson.

16 Thomas, Edward (1909). *The South Country.* Dent (republished by Little Toller books, 2009).

17 Riding, James (2016). 'A geographical biography of a nature writer'. *Cultural Geographies*, 23(3), 387–399.

18 https://concreteislands.com/the-trains-have-gone-gilroy-mere-interview/

19 https://www.caughtbytheriver.net/2020/07/the-presence-of-absence/

20 Daniels, Stephen (2023). 'Marxism, culture, and the duplicity of landscape'. In Peet, Richard & Thrift, Nigel (eds), *New Models in Geography-vol two.* Routledge.

21 https://hauntedgeneration.co.uk/2019/09/28/vic-mars-inner-roads-and-outer-paths-and-alfred-watkins-old-straight-track/

22 https://www.caughtbytheriver.net/2019/10/an-interview-with-vic-mars/

23 Thurgill, James (2015). 'A strange cartography: Leylines, landscape and "deep mapping" in the works of Alfred Watkins'. *Humanities*, 4(4), 637–652.

24 Daniels, Stephen (2006). *Lines of sight: Alfred Watkins, photography and topography in early twentieth-century Britain.* Tate Papers online.

25 https://concreteislands.com/more-like-a-school-orchestra-vic-mars-interview/

26 https://vicmars.bandcamp.com/album/inner-roads-and-outer-paths-2

27 https://concreteislands.com/clay-pipe-music-from-the-foreshore/

Chapter 4

1 See Lees, L. & Hubbard, P. (2022). '"So, don't you want us here no more?" Slow violence, frustrated hope, and racialized struggle on London's council estates'. *Housing, Theory and Society*, 39(3), 341–358.
2 https://www.versobooks.com/en-gb/blogs/news/3050-always-yearning-for-the-time-that-just-eluded-us
3 Bracewell, Michael (2021). *Souvenir*. Orion.
4 Abercrombie, P. & Paton Watson, J. (1943). *A Plan for Plymouth*. Underhill.
5 See Kolinsky, H. (2016). *Ernesettle: Everyday life in 'a lovely estate': Post-war council housing and cultural incorporation 1950–1980*. PhD thesis, University of Arts London.
6 Essex, Stephen & Brayshay, Mark (2016). 'Planning the reconstruction of war-damaged Plymouth, 1941–1961: Devising and defending a modernisation agenda'. In Clapson, Mark & Larkham, Peter (eds), *The blitz and its legacy*. Routledge, 165–180.
7 https://player.bfi.org.uk/free/film/watch-alice-in-plymouth-civic-centre-wonderland-1962-online
8 See https://www.youtube.com/watch?v=hVaxfWaIYZs
9 Gordon Chapman-Fox quoted in interview with Neil Mason, *Moonbuilding*. Spring 2023, https://moonbuilding.bandcamp.com/album/moonbuilding-sprung-2023
10 This was associated with the satellite New Towns around London, especially Harlow, nicknamed 'Pram Town'. See Llewellyan, Mark (2004). 'Producing and experiencing Harlow: Neighbourhood units and narratives of New Town life 1947–53'. *Planning Perspectives*, 19(2), 155–174.
11 Press release cited at https://warrington-runcorn-cis.bandcamp.com/album/the-nations-most-central-location
12 https://thequietus.com/articles/32942-warrington-runcorn-new-town-development-plan-interview
13 Gordon Chapman-Fox interviewed in Wicks, Simon (2023). 'Paradigms lost'. *Planner*, April 2022, 17–20, https://www.theplanner.co.uk/2022/03/22/paradigms-lost-interview-gordon-chapman-fox
14 Thoburn, Nick (2022). *Brutalism as Found: Housing, form and crisis at Robin Hood Gardens*. Goldsmith.
15 Hugh (2010) 'What went wrong at Runcorn' https://www.architectmagazine.com/design/urbanism-planning/what-went-wrong-at-runcorn_o.
16 Dellaria, Salvatore (2023). *The housing of politics in racial capitalist Britain: Merseyside, the Runcorn new town, and the Southgate Estate, 1955–1992*. Doctoral dissertation, Cornell University.
17 Gordon Chapman-Fox interviewed in Wicks, 'Paradigms Lost'.

18 Hatherley, Owen (2009). *Militant Modernism*. John Hunt, 45.
19 Gordon Chapman-Fox quoted in Fisher, Robert (2024). 'Care in the community'. *Electronic Sound,* 16 May, https://www.electronicsound.co.uk/features/long-reads/warrington-runcorn-new-town-development-plancare-in-the-community/
20 Gordon Chapman-Fox quoted in interview with Neil Mason, *Moonbuilding* 19a, 24 May, https://moonbuilding.substack.com/p/issue-19a-24-may-2024
21 https://hauntedgeneration.co.uk/2023/04/12/warrington-runcorn-new-town-development-plan-gordon-chapman-fox-and-chock-a-block/
22 https://player.bfi.org.uk/free/film/watch-prospect-of-skelmersdale-1971-online
23 Cottrell Boyce, Frank (2016). 'Ballads for Lancashire's lost utopia'. *The Observer,* 22 May, https://www.theguardian.com/society/2016/may/22/prospect-of-skelmersdale-magnetic-north-housing-new-towns
24 Fisher, M. (2012). 'What is hauntology?' *Film Quarterly*, 66(1), 16–24.
25 Mould, Oli (2017). 'Brutalism redux: Relational monumentality and the urban politics of brutalist architecture'. *Antipode*, 49(3), 701–720.

Chapter 5

1 Shoard, Marion (2002). 'Edgelands'. In Jenkins, Jennifer (ed.), *Remaking the Landscape: The changing face of Britain*. Profile Books, 117–146. This is an important work in the attempt to reassess England's liminal spaces and one that has done much to highlight their role as spaces of accidental ecological richness.
2 Houghton, Frank & Houghton, Sharon (2015). 'Therapeutic micro-environments in the Edgelands: A thematic analysis of Richard Mabey's *The Unofficial Countryside*'. *Social Science & Medicine*, 133, 280–286.
3 See Vimal, Ruppert (2022). 'The impact of the COVID-19 lockdown on the human experience of nature'. *Science of the Total Environment*, 803, 149571.
4 Kanngieser, A. M. (2015). 'Geopolitics and the Anthropocene: Five propositions for sound'. *GeoHumanities*, 1(1), 80–85.
5 For example, *Lockdown Patchwork* by folk electronic musician John Burton (released under the name Leafcutter John) sampled from thirteen hours of contributed field recordings (by E. L. Heath, Sam Tucker, and others) alongside interviews to stress the value of green spaces and allotments in the city; see https://soundcloud.com/leafcutterjohn/lockdown-patchwork. See also Schafer, Murray (1993). *The soundscape: Our sonic environment and the tuning of the world*. Simon and Schuster.

6 Wardle Armstong (2020). *Ryebanks Felds: Archaeological desk-based assessment*, https://www.saveryebankfields.co.uk/wp-content/uploads/2023/01/wardellarmstrong-archaeological-report.pdf
7 Cited in Ashton, Jenny (2022). *A working research report on Ryebank Fields*, https://www.saveryebankfields.co.uk/wp-content/uploads/2022/10/a-working-research-report-on-ryebank-fields-v1.2-jenna-ashton-2020.pdf
8 https://www.huckmag.com/article/the-protestors-battling-to-save-ryebank-fields
9 Ward, Colin (1978). *The Child in the City*. Penguin.
10 Flynn, H. (2015). 'Pomona: The lost island of Manchester'. *The Guardian*, 7 August, https://www.theguardian.com/cities/2014/aug/07/pomona-lost-island-manchester-dockland-wasteland-oasis
11 Suviste H. (2019). *Way Down to Pomona.* walk · listen · create. https://walklistencreate.org/walkingpiece/way-down-to-pomona/
12 Stoetzer, Bettina (2018). 'Ruderal ecologies: Rethinking nature, migration, and the urban landscape in Berlin'. *Cultural Anthropology,* 33(2), 295–323.
13 Jones, Lynsey (2020). 'How to tap into creativity in lockdown'. *The Financial Times*, 22 April, https://www.ft.com/content/75e4231a-7a5c-11ea-af44-daa3def9ae03
14 Harwood, Elain (2014). *Space, Hope, and Brutalism: English architecture, 1945–1975*. Yale University Press.
15 https://atthebarrier.com/2023/06/01/liz-hanks-land-album-review/
16 On the value of the edgelands, see especially Farley, Paul & Symons-Roberts, Michael (2012). *Edgelands*. Jonathan Cape. Other recent works that have an interest in such quotidian landscapes include Cowen, Rob (2015). *Common ground.* Hutchinson; Rees, Gareth (2013). *Marshland: Dreams and nightmares on the edge of London.* Influx; Woolfson, Esther (2015). *Field notes from a hidden city.* Granta; and Harrison, Melissa (2013). *Clay*. Bloomsbury.
17 See Moran, Joe (2014). 'A cultural history of the new nature writing'. *Literature & History,* 23(1), 49–63.
18 See Pryor, Melanie (2017). 'Eco-autobiography: Writing self through place'. *A/B: Auto/Biography Studies,* 32(2), 391–393. There are clear parallels here between the new nature writing and more established traditions of psychogeographically inflected urban writing: see Bonnett, Alastair (2009). 'The dilemmas of radical nostalgia in British psychogeography'. *Theory, Culture & Society,* 26(1), 45–70.
19 The label set up by Portal's Scott Sinfield, who later recorded a single EP as Ringinglow, named after a Peak District village.
20 Ben Holton cited at https://wyrddaze.wordpress.com/2024/02/11/wyrd-question-daze-epic45/
21 The Kelvedon Hatch nuclear bunker in Essex is an important underground performance space that has hosted hauntological and experimental music on

several occasions. See also Wilson, Louise (2020). 'Sounds from the bunker: Aural culture and the remainder of the cold war'. *Journal of War & Culture Studies*, 13(1), 33–53.

22 https://opus.ing/reviews/england-fallen-over-ep-epic45

23 Notably, the label also released Richard Youngs' *Memory Ain't No Decay* (2018), which was guided by this list of concerns.

24 See, for example, Simon Scott's album *Below See Level* (2012), *FloodLines* (2016) and *Long Drove* (2023).

25 Waites, Ian (2013). '"Places were I forgot things": Memory, identity, and the British council estate in the paintings of George Shaw'. *Cultural Politics*, 9(3), 357–370.

26 Dorling, Dan (2019). 'Dying quietly: English suburbs and the stiff upper lip'. *The Political Quarterly*, 90(1), 32–43.

27 Cited in Holten, Ben (2024). 'Remnants – the suburb as edgeland'. *Landnotes*, 1(Autumn), https://waysideandwoodlandrecordings.bandcamp.com/album/landnotes-1

Chapter 6

1 On the politics of psychogeography see Sidaway, James (2022). 'Psychogeography: Walking through strategy, nature and narrative'. *Progress in Human Geography*, 46(2), 549–574.

2 https://www.nativve.com/people/an-interview-with-alex-hornsby-from-weird-walk/

3 Pitcher, Ben (2022). *Back to the Stone Age: Race and prehistory in contemporary culture*. McGill-Queen's Press.

4 Barclay, Gordon & Brophy, Ken (2021). '"A veritable chauvinism of prehistory": Nationalist prehistories and the "British" late Neolithic mythos'. *Archaeological Journal*, 178(2), 330–360.

5 https://www.resurgence.org/magazine/article6135-stone-club.html

6 https://uncannylandscapes.podbean.com/e/uncanny-landscapes-16-stone-club/

7 Cook, Gary (2017). 'The arts interview: Soundscape artist Matthew Shaw', https://theecologist.org/2017/mar/23/arts-interview-soundscape-artist-matthew-shaw

8 https://www.caughtbytheriver.net/2024/02/ancient-pathways-and-modern-roads/

9 Matless, David (2024). *England's Green: Nature and culture since the 1960s*. Reaktion.

10 English, Penny (2002). 'Disputing Stonehenge: Law and access to a national symbol'. *Entertainment Law*, 1(2), 1–22.

11 Harris, Alexandra (2023). *Romantic Moderns: English writers, artists and the imagination from Virginia Woolf to John Piper*. Thames & Hudson.

12 Smith, Phil (2024). *Albion's Eco-eerie: TV and movies and the Haunted Generation.* Temporal Boundary.

13 McNaughton, Douglas (2023). 'The pattern under the plough: Folk horror in 1970s British children's television'. In Edgar, Robert & Johnson, Wayne (eds), *The Routledge Companion to Folk Horror*. Routledge, 204–217.

14 Edensor, Tim & Brophy, Kenneth (2023). 'The potent urban prehistory of an ancient megalith: The Kempock Stone, Gourock, Scotland'. *International Journal of Heritage Studies*, 29(1–2), 81–96.

15 Edensor, Tim (2017). 'Seeing with light and landscape: A walk around Stanton Moor'. *Landscape Research*, 42(6), 616–633.

16 https://thequietus.com/interviews/kemper-norton-interview/

17 Till, Rupert (2023). 'Presenting Archaeoacoustics results using multimedia and VR technologies'. *Open Archaeology*, 9(1), 20220340.

18 Knapton, Sarah (2019) '"Neolithic Brexit" unearthed at Stonehenge shows British identity began 5,000 years ago, archaeologists say'. *The Daily Telegraph*, 14 March, https://www.telegraph.co.uk/science/2019/03/13/british-identity-began-5000-years-ago-late-neolithic-brexit/

19 https://www.sapiens.org/archaeology/stonehenge-nationalism/

20 Muir, Tom & Richards, Colin (2013). 'Constructing through discourse: The folklore of stone circles and standing stones in the Northern and Western Isles'. In Richards, Colin (ed.), *Building the Great Stone Circles of the North*. Windgather, 292–300.

21 Nolan, Claire (2019). 'Sites of existential relatedness: Findings from phenomenological research at Stonehenge, Avebury and the Vale of Pewsey, Wiltshire, UK'. *Public Archaeology*, 18(1), 28–51.

Chapter 7

1 Fisher, Mark (2017). *The Weird and the Eerie*. Watkins, 61.

2 Ibid., 15.

3 Mieville, C. (2008). 'James M.R. and the quantum vampire: Weird; hauntological: Versus and/or and and/or or?' *Collapse, IV*, 105–128.

4 Luckhurst, Roger (2022). 'Brexitland's dark ecologies: New British landscape writing'. *Textual Practice*, 36(5), 711–731.

5 Tyree, J. M. (2023). 'Specters of Brexit in recent British horror'. *Film Quarterly*, 77(2), 10–19.

6 Hall, Tim (2020). 'After the rural idyll: Representations of the British countryside as a non-idyllic environment'. *Geography*, 105(1), 6–17.

7 Galea, Alexa (2014). 'Fear of folk: Why folk art and ritual horrifies in Britain'. *Journal of Illustration*, 1(1), 77–100.

8 See https://ayearinthecountry.co.uk/

9 Thurgill, James (2020). 'A fear of the folk: On topophobia and the horror of rural landscapes'. *Revenant: Critical and Creative Studies of the Supernatural*, 5, 33–56.

10 Greenaway, Jon (2024). *Capitalism, a Horror Story: Gothic marxism and the dark side of the radical imagination.* Repeater.

11 Holloway, Julian (2023). 'Sounding folk horror and the strange rural'. In Edgar, R. & Johnson, W. (eds), *The Routledge Companion to Folk Horror*. Routledge, 286–295.

12 https://www.visitmanchester.com/ideas-and-inspiration/blog/read/2019/01/manchester-folk-horror-festival-2-an-interview-with-panel-chair-julian-holloway-b729

13 Edgar, Robert (2023). 'Hinterlands and SPAs'. In Edgar, R. & Johnson, W. (eds), *The Routledge Companion to Folk Horror* . Routledge, 365–379.

14 https://folkhorrorrevival.com/2020/07/21/rural-eerie-new-album-from-flange-circus-2/

15 Hopper, Justin (2017). *The Old Weird Albion*. Penned in the Margins.

16 Sharp, Andy (2020). *The English Heretic Collection: Ritual histories, magickal geography*. Watkins.

17 This is a traditional dance used to mark off time in the calendrical year.

18 Brandon, Peter (2023). *Sussex Writers in their Landscape: Self-fulfilment in the age of the machine.* The History Press.

19 Cited in Rogers, Jude (2015). 'Shirley Collins: When I sing I feel past generations standing behind me'. *The Observer*, 31 May, https://www.theguardian.com/music/2015/may/31/shirley-collins-sing-past-generations-standing-folk-music

20 Cited in Anon (2017). 'Shirley Collins: My coast'. *Coast Magazine*, 20 May, https://www.coastmagazine.co.uk/people/coast-celebrities/my-coast-shirley-collins/

21 Jupp, Jim (2023). Coted in interview with Justin Hopper. *Uncanny Landscapes*, Podcast no. 19, https://uncannylandscapes.podbean.com/e/uncanny-landscapes-19-the-path-with-jim-jupp-of-ghost-box-records/

22 https://aquariumdrunkard.com/2024/05/08/transmissions-jim-jupp-belbury-poly-ghost-box-records/

23 Jupp, Jim (2023). Cited in interview with Justin Hopper.

24 https://hauntedgeneration.co.uk/2020/08/07/belbury-poly-jim-jupp-and-ghost-box-records/

25 Jupp, Jim (2023). Cited in interview with Justin Hopper.

26 Larrington, Carolyne & Hield, Fay (2021). 'Making "modern fairies": Making fairies modern'. *Folklore*, 132(1), 72–96.
27 Bennett, Jane (2001). *The Enchantment of Modern Life: Attachments, crossings, and ethics*. Princeton University Press.
28 Woodyer, Tara & Hilary Geoghegan (2013). '(Re) enchanting geography? The nature of being critical and the character of critique in human geography'. *Progress in Human Geography*, 37(2), 195–214. See also: Norman-Mason, Jon (2024). 'Myth-Making and the urban: Alienation, folklore and (re-) enchanting the land'. In Holloway, Phillipa & Jordan-Baker, Craig (eds), *Writing Landscape and Setting in the Anthropocene: Britain and beyond*. Springer, 49–69.
29 https://bonfirehill.bandcamp.com/album/a-year-on-bonfire-hill-january
30 See http://www.stormchorus.co.uk/about.php
31 See Simons, Matt (2021). 'Embodied Englishness in the inter-war Morris Revival'. In Cheesman, Matthew & Hart, Carina (eds), *Folklore and Nation in Britain and Ireland*. Routledge, 132–145.
32 https://thevinylfactory.com/features/the-archivist-audio-visual-myth-making-with-meticulous-independent-label-folklore-tapes/
33 Parr, George (2023). 'The spectre of trauma in the myth of Psychick Albion'. *Undefined Boundary*, 1(2), 71–86.

Chapter 8

1 The track was later revealed as 'May Birching', referring to the largely forgotten ritual of decorating homes with birch twigs and flowers on the first day of Spring, part of a ritual thought to promote fertility and well-being. This was later released on a limited-edition tape sold at the Thwaites Mill gig, September 2023.
2 See especially Hauser, Kitty (2007). *Shadow Sites: Photography, archaeology, and the British landscape 1927–1955*. Oxford University Press.
3 Phillips, Martin, Smith, Darren, Brooking, Hannah, & Duer, Mara (2021). 'The agencies of landscape in rural gentrification: Impressions from the wood, the village and the moortop'. *Sociologia Ruralis*, 61(4), 778–807.
4 Spracklen, Karl (2018). 'Theorising northernness and northern culture: The north of England, northern Englishness, and sympathetic magic'. In Riches, Gabby, Spracklen, Karl, & Swain, Spencer (eds), *Northernness, Northern culture and Northern narratives*. Routledge, 4–16.
5 Hoskins, W. G. (1955). *The Making of the English Landscape*. Hodder & Stoughton.
6 White, Robert (2003). 'Founders: Arthur Raistrick'. *Landscapes*, 4(2), 111–123.
7 Raistrick, Arthur (1946). *The Story of the Pennine Walls*. Dalesman.

8 Myers, Ben (2022). 'Introduction'. In Hughes, Glyn, *Milstone Grit.* Little Toller Books. See also https://www.caughtbytheriver.net/2022/08/millstone-grit-benjamin-myers-introduction/
9 Hughes, Glyn (1987) *Millstone Grit.* Pan.
10 Skelton, Richard (2020). *North of Here: Imagining the human and other-than-human in late-upper Palaeolithic Britain.* Doctoral thesis, Manchester Metropolitan University.
11 Macfarlane, Robert (2019). *Underland.* Penguin.
12 Overend, David, Lorimer, Jamie, & Schreve, Danielle (2020). 'The bones beneath the streets: Drifting through London's Quaternary'. *cultural geographies*, 27(3), 453–475.
13 https://louderthanwar.com/interview-spaceship-talk-ambient-psychogeology-yorkshire/
14 'Homunculus at Staup's Mill' is a track by Hebden Bridge-based performer Colin Robinson (on the album 202 Days), who records as Jumble Hole Clough, suggesting a particular density of interest in the post-industrial landscapes of the Calder Valley. Both Hebden and Todmorden are gentrified communities known as hubs for counter-cultural performance and creativity, especially around the wyrd and hauntological.
15 Palsson, G. & Swanson, H. A. (2016). 'Down to earth: Geosocialities and geopolitics'. *Environmental Humanities*, 8(2), 149–171.
16 Clark, Nigel (2017). 'Politics of strata'. *Theory, Culture & Society*, 34(2–3), 211–231.
17 Colquhoun, Ithell (2017). *The Lliving Stones: Cornwall.* Peter Owen, 3.
18 Bosworth, Kai (2024). 'Filling the hole? On new geographies of the subsurface'. *Progress in Human Geography*, 48(2), 151–171.

Chapter 9

1 Blum, Hester (2013). 'Introduction: Oceanic studies'. *Atlantic Studies*, 10(2), 151–155.
2 Shields, Rob (1991). *Places on the Margin.* Routledge.
3 House of Lords (2019). *Select committee on regenerating seaside towns and communities the future of seaside towns.* HL Paper 320, 4 April, https://publications.parliament.uk/pa/ld201719/ldselect/ldseaside/320/32002.htm
4 Burdsey, Daniel (2016). *Race, Place and the Seaside: Postcards from the edge.* Springer.

5 Sykes, Tom & Netter, Louis (2024). 'The littoral and the metaphorical: Psychogeography, liminality and English seaside towns'. In Morrison, James (ed.), *Silenced voices and the media: Who gets to speak?* Routledge, 213–228.
6 Bolton, Tom (2018). *Low Country: Brexit on the Essex Coast.* Penned in the Margins.
7 Bunting, Madeleine (2023). *The Seaside: England's love affair.* Granta.
8 Le Brun, Lily (2022). *Looking to Sea.* Sceptre.
9 On Margate's putative middle-class renaissance, see Hubbard, Phil (2017). *The Battle for the High Street: Retail gentrification, class and disgust.* Palgrave.
10 Burdsey, *Race, place and the seaside.*
11 Prosser, Bethan (2022). *Listening to Urban Seaside Gentrification: Living with displacement injustices on the UK south coast.* Doctoral thesis, University of Brighton.
12 Willett, Joanie (2023). 'Counter-urbanisation and a politics of place: A coastal community in Cornwall and rural gentrification'. *Habitat International*, 141, 102935.
13 Ash, Lamorna (2019). 'Coastal erosion: Modern life in a Cornish fishing village'. *TLS. Times Literary Supplement*, 6077, 20–21.
14 Arnall, A. (2023). 'Encountering the Anthropocene: Reconfiguring human-nature relations on the North Norfolk Coast, UK'. *Geoforum, 143*, 103768.
15 See Matless, David (2017). 'The anthroposcenic'. *Transactions of the Institute of British Geographers*, 42(3), 363–376; Matless, David (2018). 'The anthroposcenic: Landscape in the anthroposcene'. *British Art Studies*, 10; Matless, David (2018). 'Next the sea: Eccles and the anthroposcenic'. *Journal of Historical Geography*, 62, 71–84.
16 Arnall, 'Encountering the Anthropocene', 103768.
17 Woollett, Lisa (2024). *Lost to the Sea.* John Murray.
18 Rees, Gareth (2023). *Sunken Lands.* Elliot and Thompson.
19 Farrer, Adam (2022). *Cold Fish Soup.* Saraband, 25.
20 Davis, Sophie (2020). *Island Thinking: Suffolk stories of landscape, militarisation and identity.* Springer.
21 Sebald, W. G. (2013). *The Rings of Saturn.* Vintage Classics, 237.
22 Owens, Susan (2020). *Spirit of Place: Artists, writers and the British landscape.* Thames & Hudson.
23 https://irp.blogs.bristol.ac.uk/2019/02/11/researching-secrecy-amidst-the-ruins-of-orford-ness/
24 See Mulholland, Drew (2024). *Elizabethan Tape Loops.* Hidden Britain Books.
25 See https://www.electronicsound.co.uk/features/long-reads/drew-mulholland-tales-from-the-loop/

26 Skrimshire, Stefan (2021). 'The view from Dunwich'. *Environmental Humanities*, 13(1), 264–271.
27 Sebald, *The Rings of Saturn*, 159.
28 Skrimshire, 'The view from Dunwich', 264–271.
29 See Cassidy, Kathryn, Yuval-Davis, Nira, & Wemyss, Georgie (2018). 'Debordering and everyday (re) bordering in and of Dover: Post-borderland borderscapes'. *Political Geography*, 66, 171–179.
30 Dillon, Brian (2003). 'Listening for the enemy'. *Cabinet Magazine*, Fall/Winter 12, https://www.cabinetmagazine.org/issues/12/dillon.php
31 https://brachliegentapes.bandcamp.com/album/do-not-lose-sight-of-the-sea
32 https://brachliegentapes.bandcamp.com/album/leadlined

Chapter 10

1 McPherson, Charlotte (2024). 'Considering the past, present and future: Making the case for a hauntological approach to youth'. *Journal of Youth Studies*, https://doi.org/10.1080/13676261.2024.2328121
2 Boym, Svetlana (2002). *The Future of Nostalgia*. Basic Books. See also Bonnett, Alastair (2015). *The Geography of Nostalgia: Global and local perspectives on modernity and loss*. Routledge; and Legg, Steve (2004). 'Memory and nostalgia'. *cultural geographies*, 11(1), 99–107.
3 Luckhurst, Roger (2022). 'Brexitland's dark ecologies: New British landscape writing'. *Textual Practice*, 36(5), 711–731.
4 Fisher, Alice (2023). 'Cool as folk: Why Britain's young rebels are embracing ancient rites'. *The Guardian*, February 12, https://www.theguardian.com/culture/2023/feb/12/folk-customs-britain-young-generation
5 Fisher, Mark (2014). *Ghosts of My Life*. Zero Books, 28.

Select Discography

The Advisory Circle *As the Crow Flies* Ghost Box Records GBX015 and GBX015LP 2011, CD and 12-inch vinyl

The Advisory Circle *Full Circle* Ghost Box Records GBX042 and GBX042LP 2022, CD and 2 x 10-inch vinyl

Ætheling *West Dorset* self-released 2023, download only

The Assembled Minds *Psychogeographic Audio Explorations Along the A303 Vol.1* Patterned Air, AIR006 2019, lathe-cut 8 inch vinyl

Astley, Virginia *From Gardens Where We Feel Secure* Happy Valley Records/Rough Trade ROUGH 58 1983, 12 inch vinyl, RETRADECD001 2003, CD

Astley, Virginia *Hope in A Darkened Heart* WEA 159 112-1 1986, 12-inch vinyl

Astley, Virginia *Maiden Newton Ecliptic* self-released 2021, download only

Astley, Virginia *The Singing Places* self-released 2023, download only

The Balloonist *The Balloonist* Wayside and Woodland W&W058 2022, CD

The Balloonist *A Quiet Day* Wayside and Woodland W&W070 2024, limited CD

Belbury Poly *The Owl's Map* Ghost Box Records GBX007 2006, CD and 12-inch vinyl

Belbury Poly *Belbury Tales* Ghost Box Records GBX016 2012, CD and 12-inch vinyl

Belbury Poly *The Gone Away* Ghost Box Records GBX035 2021, CD and 12-inch vinyl

Belbury Poly *The Path* Ghost Box Records GBX043 2023, CD and 12-inch vinyl

John Biddulph *Doll Tor* self-released 2021, download only

John Biddulph *Perambulations I* self-released 2023, CD (edition of 100)

Birds in the Brickwork *Twelve Months* Wayside and Woodland 053 2021, FLAC download and accompanying calendar

Birds in the Brickwork *Recovery* Wayside and Woodland 055 2022, CD and photobooklet (100 copies)

Birds in the Brickwork *A Strange Place* Wayside and Woodland 064 2023, CD and photobooklet (15 copies)

Black Zephyr *M6 Travelogue* Murmurations M03 2023, lathe cut 7-inch, edition of 49

Bloxom Research Group *Zener_42* self-released 2023, CD and cassette with field notes

Bonfire Hill *A Year on Bonfire Hill* self-released, monthly download 2023

Bonfire Hill *A Year on Bonfire Hill: Sendelica Remix* Bonfire Hill Records BON05 2024, 12-inch vinyl

David Boulter *Yarmouth* Clay Pipe Music pipe_026 and pipe_026cass 2020, 12-inch coloured yellow vinyl and cassette, pipe_026_02 and pipe_026_CD 2021, 12-inch blue vinyl and cassette

David Boulter *Lover's Walk* The Spoken Word 1 2021, 12-inch green vinyl (also available in limited edition 12-inch vinyl instrumental version, The Spoken Word 2 2022, 250 copies)
David Boulter *St Ann's* Clay Pipe Music pipe_039, 2024, CD and 12-inch purple vinyl
Brave Timbers *Hope* Little Cracked Rabbit LC007 2016, CD
Rob Bridgett *Dungeness Array* self-released 2013, download only
Broadcast *Accidentals* Wurlitzer Jukebox WJ06 1996, 7-inch vinyl
Broadcast *The Noise Made By People* Warp WARPCD65 and WARPLP65 2010, CD and 12-inch vinyl
Broadcast *Berberian Sound Studio* Warp WARPCD233 and WARPLP233 2013, CD and 12-inch vinyl
Broadcast and the Focus Group *Familiar Shapes and Noises* Ghost Box GBX704 2010, 7-inch vinyl
Broadcast and the Focus Group *Investigate Witch Cults of the Radio Age* Warp WARPCD189 and WARPLP189 2009, CD and 12-inch vinyl
Jon Brooks *Autres Directions* Clay Pipe Music pipe_015 2017 mist coloured vinyl (500 copies), and pipe_015-2 2017, limited edition repress orange 12-inch (500 copies)
Jon Brooks *Shapwick* Clay Pipe Music pipe_004 2012, CDr, pipe_07 2013, 12-inch coloured vinyl, pipe_07_03 2020, 12-inch clear blue vinyl repress
Jon Brooks *Walberswick* More Than Human MTH007 2015, 12-inch vinyl
Will Burns and Hannah Peel *Chalk Hill Blue* Rivertones RIVERTONE10 and RIVERTONE10CD 2018, 12-inch vinyl and CD
Cargill, Alexander R. Esq *Bromham* Woodford Halse WF89 2024, limited CD
Carlisle City Council *The Lanes Re-development* Subexotic SUBEX000142 2024, 12-inch clear vinyl, limited edition of 150
Carlisle City Council *Hardwicke Circus* self-released 2024, download only
Cautionary Guides *A Cautionary Drive Around the Wirral* Castles in Space CiS-SL040 2023, download and PDF booklet
The Central Office of Information *The Central Office of Information* Castles in Space CiS031 2019, limited CD
The Central Office of Information *Private Issue Synthetic Music 2011-2020* Woodford Halse WF19 2021, limited CD
Iain Chambers *Secrets of Orford Ness* self-released 2020, download only
Gordon Chapman Fox *Nine Travellers* Castles in Space CiS-SL039 2023, yellow transparent 12-inch vinyl
David Chatton Barker and Ian Humberstone *Megalithic Monuments Volume One – Stanton Drew Stone Circle* Folklore Tapes FTMMI 2016, numbered limited edition 10-inch vinyl
David Chatton Barker and the Whitworth and Healey Vale Brass Band *Brown Wardle Hill* Lancashire Folklore Tapes LFTVII 2023, 12-inch vinyl

Dave Clarkson *A Red Guide to the Coastal Quicksands of the British Isles* Linear Obsessional Recordings LOR087 2017, CDr (edition of 100)

Dave Clarkson *A Blue Guide to Shore Ghosts and Sea Mystery* Linear Obsessional Recordings LOR099 2017, CDr (edition of 100)

Dave Clarkson *A Pocket Guide to Subterranea: mysterious caves of the British Isles* Linear Obsessional Recordings LOR134 2019, cassette

Dave Clarkson *A Pocket Guide to Wilderness: deep forests and dark woods of the British Isles* Linear Obsessional Recordings LOR159 2021, cassette (two editions of 50 in different colour shell)

Dave Clarkson *A Pocket Guide to Dream Land (Faded Fairgrounds and Coastal Ghost Towns of the British Isles)* Cavendish House CAVcass02 and CAVCD02 2022, cassette (edition of 120) and CD (edition of 300)

Dave Clarkson *A Pocket Guide to Further Afield Recordings (Obscure Locations of the British Isles)* Cavendish House CAVCD03 2023, CD and poster set (edition of 100) and standard CD (edition of 300)

Clevelode *Muntjac* Fenny Compton FC06 2024, CD

Conflux Coldwell *The Phantomatic Coast* Subexotic Records SUBEX00068 2022, 12-inch vinyl

Conflux Coldwell *The Sunshine Miners* Woodford Halse WF93 2025, cassette

Shirley Collins *Lodestar* Domino WIGLP389 2016, 12-inch vinyl and CD

Shirley Collins *Heart's Ease* Domino WIGLP454 2020, 12-inch vinyl and CD

Shirley Collins *Archangel Hill* Domino WIGLP494 2023, 12-inch vinyl and CD

Concretism *Town Planning* Public House Recordings Flat 006 2014, 12-inch vinyl

Concretism *The Thetford Beast* Castles in Space CiS091 2022, 10-inch vinyl

Concretism *39 Furnival Street* self-released 2024, download only

Craven Faults *Netherfield Works* Lowfield Works Parcel 322 2017, 12-inch vinyl

Craven Faults *Springfield Works* Lowfield Works Parcel 369 2019, 12-inch vinyl

Craven Faults *Nunroyd Works* Lowfield Works Parcel 529 2019, 12-inch vinyl

Craven Faults *Lowfold Works* Lowfield Works Parcel 731 2019, 12-inch vinyl

Craven Faults *Enclosures* Leaf DOCK 73 2020, 12-inch vinyl

Craven Faults *Erratics and Unconformities* Leaf DOCK 73 2020, CD and 2 x 12-inch vinyl

Craven Faults *Standers* Leaf BAY131 2023, CD and 12-inch vinyl

Craven Faults *May Birching* Lowfold Works Parcel451 2023, cassette, edition of 100

Craven Faults *Thwaite Watermill* Lowfold Works Parcel46 2024, cassette, edition of 100

Craven Faults *Bounds* mini-LP Leaf BAY140X and BAY140VX 2024, 12-inch vinyl and limited editions in 'Shake Holl Green', 'Lime Kiln Grey' and 'Tarn Silk Blue' coloured vinyl

Amy Cutler *The Ends (also End) of (the) Earth and Variants* Crow Versus Crow CV021 2021, cassette

The Declining Winter *Endless Scenery* Sound in Silence sis027 2015, CDr

The Declining Winter *Belmont Slope* Home Assembly HAM018 2018, CDr, 12-inch vinyl
The Declining Winter *Recordings of Weird Air* Signal Records S007 2021, cassette
The Declining Winter *Really Early, Really Late* Home Assembly HAM030 2023, CD, 2 x 12-inch vinyl and cassette
Degradation *Leadlined* Brechliegan Tapes 008 2022, 12-inch vinyl
Deliquescent Crystals *Time Turns into Space* Castles in Space CiS114 2022, CD
Dollboy *Ghost Stations / Geisterbahnhöfe* Second Language SL026 2013, CD with 3-inch bonus CD (SL026x) of remixes (*Sidings* EP)
EL Heath *Snailbeach Mines Trust* Wayside and Woodland Recordings W&W009 2010, CD
EL Heath *Shropshire Hill Country* Wayside and Woodland Recordings W&W007 2010, CD
EL Heath *A Song for the Village of New Invention* Plenty Wenlock PW002 2012, mini-CD
EL Heath *Companions* Plenty Wenlock PW010 2023, CDr (edition of 20)
Broken English Club The English Beach L.I.E.S. LIES-094 2017, double 12-inch vinyl
epic45 *Against the Pull of Autumn* Where are My Records WAM015 2004, CD
epic45 *Long Walks EP* self-released 2005, limited edition numbered CDr
epic45 *Drakelow EP* Make Mine Music MMM0117 2006, CD and 12-inch vinyl
epic45 *May Your Heart be the Map* Make Mine Music MMM013 2007, CD
epic45 *In All the Empty Houses* Make Mine Music MMM047 2009, CD
epic45 *Weathering* Make Mine Music MMM068 2011, Wayside and Woodland Recordings W&W052 10th year anniversary edition 2021, 2 x 12-inch transparent vinyl
epic45 *Through Broken Summer* Wayside and Woodland Recordings W&W 2018, CD and an all-media edition CD, cassette, 2 x 12-inch vinyl
epic45 *We Were Never Here* Wayside and Woodland Recordings W&W048 2020, CD
epic45 *Cropping the Aftermath* Wayside and Woodland Recordings W&W049 2020, CD and 12-inch blue vinyl
epic45 *You'll Only See Us When the Light Has Gone* Wayside and Woodland Recordings W&W049 2023, CD and 12-inch vinyl
The Falcons of Haunt *Sometimes a Spectre, always a King* self-released 2024, download only
Flange Circus *Rural Eerie* self-released 2020, DVD
Flange Circus *Katabasis* Analog Trash Records AT0164 2023, CD
Flaer *Preludes* Odda Recordings ODA 01M and ODA 01C 2023, 12-inch vinyl mini-album and cassette
Flaer *Burrow* Odda Recordings ODA EP3 2024, cassette
Nancy Gaffield and the Drift *Wealden* Longbarrow Press 2010, CD with booklet

Gazelle Twin *Pastoral* Anti Ghost Moon Records AGMR036 and AGMR036CD 2018, 12-inch vinyl and CD
Ghedi, Jim *A Hymn for Ancient Land* Basin Rock BR005 2018, 12-inch vinyl and CD
Gilroy Mere *The Green Line* Clay Pipe Music pipe_016 2016, 12-inch vinyl and cassette
Gilroy Mere *Over the Tracks* Clay Pipe Music pipe_023 2020, three track 7-inch flexidisc (with cut out model station) (limited edition of 100)
Gilroy Mere *Adlestrop* Clay Pipe Music pipe_025 2020, 12-inch vinyl and cassette
Gilroy Mere *Gilden Gate* Clay Pipe Music pipe_033 2023, 12-inch vinyl
Glåsbird *Fenscapes* whitelabrecs WLR172 2025, limited 12-inch vinyl (100 copies)
Gwenno *Tresor* Heavenly HVNLP205 and HNVCD205 2022, 12-inch vinyl and CD
Liz Hanks *Land* Hudson Records HUD034LP and HUD034CD 2023, 12-inch vinyl and CD
Liz Hanks *Land Remixed* Hudson Records 2024, download only
The Hardy Tree *The Fields Lie Sleeping Beneath* Clay Pipe Music pipe_001 2010, CDr, reissued pipe_001_03 2021, numbered 12-inch vinyl
The Hardy Tree *Through Passages of Time* Clay Pipe Music pipe_014 2016, CDr reissued pipe_014_02 2017, numbered 12-inch yellow vinyl
The Hardy Tree *Common Grounds* Clay Pipe Music pipe_034 and pipe_034_CD 2022, 12-inch vinyl and CD
The Hardy Tree *The Theme to Stagdale EP* Clay Pipe Music pipe_020 2019, 7-inch flexi
The Hardy Tree *Stagdale in the Snow EP* Clay Pipe Music pipe_030 2021, 7-inch flexi
The Hardy Tree *All the Hours* Clay Pipe Music minipipe_007 2024, 3-inch CDr
The Hardy Tree *August 1975 EP* Clay Pipe Music pipe_042 2024, 7-inch flexi
Benjamin Harrison *Brutal* Castles in Space CiS102 2022, 12-inch vinyl
Haven Avenue *Haven Avenue* self-released 2016, download only
Darren Hayman & Secondary Modern *Pram Town* Track and Field HEAT51, 2012 CD.
Darren Hayman *Lido* 2012 Clay Pipe Music pipe_006, CDr, Where it's At is Where You Are wialop020, limited 12-inch vinyl and cassette, reissued Clay Pipe Music pipe_005_02 2023, limited 12-inch clear vinyl
Head Technician Ecstatic *Profane Architecture* Ecstatic ELP036 2018, 12-inch vinyl (green limited edition)
Heavy Cloud *Beyond the Last Beach House* self-released 2020, CDr
Heavy Cloud *Scree* Brechliegen Tapes 2023, cassette
The Heartwood Institute *Whispering Knights* Reverb Workshop RW364 2017, CDr (40 copies)
The Heartwood Institute *Secret Rites* Polytechnic Youth PV51 2018, 12-inch vinyl
The Heartwood Institute *Witchseason* self-released 2020, download only
The Heartwood Institute *Witchcraft Murder*, Library of the Occult LOTO 006 2021, 12-inch vinyl (300 copies on purple vinyl, 300 on blue) and cassette (50 copies)

The Heartwood Institute *Land of the Lakes* Castles in Space CiSSL014 2021, 12-inch red and black vinyl

The Heartwood Institute *Mardale* Sleep Fuse SF48 2022, CDr (50 copies)

The Heartwood Institute *Mist over Pendle* Doomshire Tapes 066 2023, cassette limited edition in purple (60 copies) and gold shell (second run)

Hidden Rivers *Where Moss Grows* Serein SERE008LP and SERE008 2015, 12-inch vinyl and CD

Ian Humberstone and David Chatton Barker *Two Witches* Devon Folklore Tapes DFT1, 2 x 10-inch vinyl, limited edition of 500

Ian Humberstone and David Chatton Barker *Theo Brown and the Folklore of Dartmoor* Folklore Tapes DFT-VI 2014, seven 7-inch vinyl singles with DVD (box set), reissued as double 12-inch vinyl (300 copies) in 2021

Karl Hyde *Edgeland* Universal 3729840 2013, CD

The Inward Circles *Belated Movements for an Unsanctioned Exhumation August 1st 1984* Corbel Stone Press CSPCD004 2016, limited edition. CDr

Isolated Community *Cold Sleet on Old Slate* Sea Blue Audio SBA29 2021, download only

Isolated Community *Pit Full of Ghosts* Sea Blue Audio SBA42 2023, download only

David A Jaycock *Music for Space Age Shopping* Subexotic SUBEX 00170 2024, 12-inch orange vinyl, limited edition of 150

Mark Jenkin *Bait* Invada INV235LP 2019 and INV235T 2023, 12-inch vinyl and cassette

Mark Jenkin *Enys Men* Invada INV289LP 2023, 12-inch red and yellow vinyl editions

July Skies *The English Cold* Make Mine Music MM009 2004, CD, reissued Caroline True Records CTRUE21 2017, 12-inch vinyl

July Skies *Where the Days Go* Make Mine Music MM021 2006, CD.

July Skies *The Weather Clock* Make Mine Music MM027 2008, CD, reissued 2016 , 12 Rivertones RIVERTONE5LP, 12-inch vinyl

Jumble Hole Clough *202 Days* self-released 2014, download only

Junkboy *Trains, Trees, Topophilia* Fretsore Records FR0005LP and FR0005CD, lathe-cut 12-inch vinyl and CD

Junkboy *Littoral States* Wayside and Woodland Recordings W&W065 2023, CDr limited edition of 25

Justin Hopper and Sharron Kraus with the Belbury Poly *Chanctonbury Rings* Ghost Box Records GBX033 2019, CD and 12-inch vinyl

Kemper Norton *Carn* Exotic Pylon EP30 2013, CD

Kemper Norton *Toll* Front and Follow F&F043 2015, promo CDr and download

Kemper Norton *Brunton Calciner* self-released 2020, download only

Kemper Norton *Oxland Cylinder* self-released 2020, download only

Kemper Norton *Estrenyon* self-released 2022, download only

Sharron Kraus *Friends and Enemies; Lovers and Strangers* Clay Pipe Music pipe_011 2015, 12-inch vinyl

Langham Research Centre *Gateshead Multistorey Car Park* Econore eco_125 2018, cassette

Low End Activist *Hostile Utopia* Sneaker Social Club SNKRLP 009, double 12-inch vinyl with insert

Low End Activist *Municipal Dreams* Sneaker Social Club SNKRLP 013, double 12-inch vinyl with photobook

Llyn Y Cwn *Twll Du* Cold Spring CSR254CD 2019, CD

Llyn Y Cwn *Dinorwic*, 2020 Cold Spring CSR286CD, CD

Llyn Y Cwn *Stratigraphy*, 2023 Manky Music, limited CDr

Llyn Y Cwn *Megaliths*, 2024 Cold Spring CSR334CD, CD and limited edition CD with hardback photobook (50 copies)

The Magnetic North *Prospect of Skemersdale* Full Time Hobby FTH225CD and FTH255LP 2016, CD and 12-inch vinyl

Suzy Mangion *Location: Gilsland* self-released 2022, download plus limited-edition booklet

Megalithic Transport Network *Phase One*, self-released 2022, cassette and download

Megalithic Transport Network *Phase Two*, self-released 2023, cassette and download

Megalithic Transport Network *Phase Three Site Survey: Excavations on Harthill Moor* self-released 2023, download only

Megalithic Transport Network *Phase Three*, self-released 2024, download only

Megalithic Transport Network *Engine Vein* Crooked Acres 2024, download only

Megalithic Transport Network *Sutton Scarsdale* Human Geography Recordings 2024 C15 cassette and download

The Memory Band *On the Chalk (Our Navigation of the line of the Downs)* Static Caravan VAN257 2013, CD

The Memory Band *Further Navigations EP* Static Caravan VAN272 2014, 10-inch vinyl

The Memory Band *A Fair Field* Static Caravan VAN305 2016, 12-inch vinyl

The Memory Band *Never the Same Way Twice* self-released 2024, download only

Memotone *Tollard* The Trilogy Tapes TTT108 2023, 12-inch vinyl

Memotone (with Chris Yates) *Illuminations Parts I-III* self-released 2023, download only

Memotone *In the Shadow of Golden Cap* (Dorset) Folklore Tapes FTCCII 2024, cassette issue of 100, b/w Arianne Churchman *The Horse Woman of Piper's Vale* (Suffolk)

My Autumn Empire *The Village Compass* Wayside and Woodland Recordings W&W006 2010, CD

Modern Nature *How to Live* Bella Union BELLA870V and BELLA870CD 2019, 12-inch vinyl and CD

Modern Nature *Nature EP* Bella Union BELLA899V 2019, 12-inch vinyl

Modern Nature *Annual* Bella Union BELLA1038V 2020, 12-inch vinyl

Modern Nature *Island of Noise* Bella Union BELLA1223V and BELLA1223CD 2022, 12-inch vinyl and CD

Mordant Music *Misinformation* Mordant Music 044 and BFIDVD0898 2010, DVD
Mordant Music *Travelogues 1-21* Mordant Music 2007-2018, download only
Drew Mulholland *The Making of Landscape* Eighth Climate ETHC-05 2016, CDr
Drew Mulholland *A Haunting Strip of Marshland* Castles in Space Subscription Library CiS053 2020, 12 inch vinyl
Drew Mulholland *A Trip to Hob Moor* Woodford Halse WF28 2021, cassette
Mark Peters *Innerland* Sonic Cathedral SCR094 2017, cassette, SCR094LP 2018, limited edition 12-inch blue and white vinyl
Mark Peters *Ambient Innerland* Sonic Cathedral SCR094X 2018, transparent ochre 12-inch vinyl
Mark Peters *Progress* EP, self-released 2024, download only
The Night Monitor *Horror of the Hexham Heads* Library of the Occult LOTO 033 2024 12 inch grey vinyl, limited edition of 600
The Night Monitor *The Black Monk of Pontefract* self-released 2024, 7 inch black vinyl, limited edition of 100
O.G. Jigg (aka Memotone) *The Land Dictates the Lay of the Stone* Earth Memory Recordings EMR001 2022, vinyl 12-inch
Tommy Perman and Rob St. John *Waters of Life* Folklore Tapes MISC1 2014 (b/w Carl Turney and Brian Campbell's 'Williamson Tunnels' as *Chthonic Cities*), limited edition cassette
Tommy Perman *Resonant Viaduct* self-released 2023, 10-inch lathe cut vinyl with limited edition print
David J. Pitt *Angelis ad Virginem* self-released 2023, download only
Plinth *Music for Smalls Lighthouse* Second Nature SL05 2010, CDr, Clay Pipe Music PIPE08 2013, 12-inch vinyl
Pram *The Last Astronaut* Kooky KOOKY004, 7-inch vinyl
Pram *The Owl Service* Domino RUG110 and RUG110CD 2000, 7-inch vinyl and CD single
Pram *North Pole Radio Station* Domino RUG WIGLP49 and WIGCD49 1998, 12-inch vinyl and CD
Pulp *We Love Life* Island CID8110 and ILPS8110 2001, 12-inch vinyl and CD
Pulselovers *Cotswold Stone* Castles in Space CiS037 2019, 12-inch vinyl
Pulselovers *Circles within Circles* Castles in Space Subexotic SUBEX00080 2022, 12-inch vinyl
Pulselovers *Northern Minimalism 3* Castles in Space CiS110 2023, 12-inch vinyl
Pye Corner Audio *Sleep Games* Ghost Box Records GBX017CD and GBX017LP 2012, CD and 12-inch vinyl
Pye Corner Audio *Stasis* Ghost Box Records GBX025CD and GBX025LP 2016, CD and 12-inch vinyl
Pye Corner Audio *Entangled Routes* Ghost Box Records GBX039CD and GBX039LP 2021, CD and 12-inch vinyl

Rise Enormity *Kinsea Sound Mirror* self-released 2024, download only

David Rothon and Oliver Cherer *Estuary English* Clay Pipe Music pipe_032 2022, limited CD (200 copies)

Rural District Lo Fi Project *Rainy Night in Alciston* self-released 2023, download only

Rural District Lo Fi Project *Theobald's Road* self-released 2023, download only

Rural District Lo Fi Project *Ghostbusting with Harry Price* self-released 2024, download only

Rob St John *Surface Tension* Patter and Process Press PPP001 2015, limited CD and CDr, reissued by Blackford Hill BH005V and BH005CD, 2021, limited eco-mix 12-inch vinyl and CD

Rob St John *Worms or Dragons (Durham)* Folklore Tapes FTCCII 2024, cassette issue of 100, b/w Powders *Crazing* (*Staffordshire*).

Luke Sanger *Global Horizontal Irradiance* Serein SERE025 2020, download only

Robin Saville *Public Flowers* Second Language SL028 2013, CD and SL028x 2013, limited CDr (reissued by Abstrakce Records, ABST 019 2021, on cassette, edition of 40)

Robin Saville *Build a Diaroma* Morr 172-LP 2020, 12-inch vinyl and Morr 172CD 2020, CD

Robin Saville *Lore* Morr 199-LP 2023, 12-inch vinyl (limited edition of 500)

Robin Saville and Oliver Cherer *Sizewell* Modern Aviation 2019, cassette (limited edition of 150)

Scholars of the Peak *The Peak* EP self-released SOTP0001 2024, download only

Scholars of the Peak *Polymorphic* self-released 2004, download only

Simon Scott *Below Sea Level* 12k 12k1071 2012, CD and limited CD with 80-page booklet

Simon Scott *Floodlines* Touch Tone 53 2016, CD

Simon Scott *Long Drove* Room 40 RM 4138CD 2023, CD

Keith Seatman *Around Folly and Down Hill* KS Audio KS003 2014, CD

Keith Seatman *Sad Old Tatty Bunting* Castles in Space CiS093 2022, CD

Keith Seatman and Douglas Powell *Broken Folk* KS Audio KS008 2018, 10-inch clear green vinyl

Jonathan Sharp *Divided Time* Castles in Space CiS030 2024, 12-inch vinyl

Matthew Shaw *Lodge Hill* self-released 2017, download only

Matthew Shaw *Pathways of Ancient Britain* self-released 2022, download only

Matthew Shaw *A Journey to the Stones* Glass Modern GLAMCD057 2024, CD

Iain Sinclair Stone Tape Shuffle Test Centre TC001 2012, 12-inch vinyl

Richard Skelton *Landings* Sustain Release SRL14 2009, limited CDr, Type TYPE055 and TYPE055V 2010, CD and 2 x 12-inch vinyl

Richard Skelton *Last Glacial Maximum* Corbel Stone Press CSPCD008 2008, CD

Slug Milk *Wardsend Cemetry* (South Yorkshire) Folklore Tapes FTCCVI 2024, cassette issue of 100, b/w David A Jaycock *Arthur Gernow* (Cornwall)

The Soulless Party *Tales from the Black Meadow Archive* self-released 2013, CDr
The Soulless Party *The Black Meadow Archive Volume One* Castles in Space CiS 041 2020, 12-inch vinyl
The Soulless Party *The Black Meadow Archive: the lost tapes* Castles in Space CiS 078 2021, CDr
Spaceship *A Transect of Shackleton Knoll* Where It's At Is Where You Are WIACD066 2018, CD (limited edition of 49)
Spaceship *Outcrops* Where It's At Is Where You Are WIALP070 2019, 12-inch vinyl (limited edition of 300)
Spaceship *Ravines* Where It's At Is Where You Are WIALP094 2022, 12-inch vinyl
Spaceship *Structures* Where It's At Is Where You Are WIALP119 2024, 12-inch vinyl
Sproatly Smith *River Wye Suite* Weirdshire 2022, 12-inch vinyl
The Stone Tapes *Avebury* Reverb Workshop RW356 2015, CDr (limited edition of 45)
Stonecirclesampler *After The Ice Age* Human Geography 2023, cassette
Stonecirclesampler *Deep Dream Derbyshire Gloom* Brachliegen Tapes 036 2024, cassette
Storm Chorus *Died for Love* Green Grass Records GGR001 2019, CD
Sub Loam *The Unfolding Map* Wist Records WistRec025 2017, 3 x CDr box set
Sub Loam *Excavated Relics* Dissolving Records Henge013 2019, cassette (limited edition of 45)
Suviste, Hayley *Edgelands* commissioned by Huddersfield Contemporary Music Festival 2021, download only
Sunshine Playroom, *The Old Railway Line* self-released 2023, CD (limited edition of 100)
Swansither *Ronsack* Subexotic Records SUBEX00176 2025, CD and limited 12" vinyl
Michael Tanner *The Sussex Workings Vol 3: music for Three Hills* self-released 2015, download only
Michael Tanner *Suite for Psaltary and Dulcimer* Kit Records KR15 2016, 12-inch vinyl
Hayden Thorpe *Ness* Domino WIGLP516XM and WIGCD156, 12-inch vinyl and CD
The Twelve Hour Foundation *The Lighter Side of Concrete* Disques Boum boum0017 2013, CD
The Twelve Hour Foundation *Six Twenty Negative* Castles in Space CiS065 2020, 12-inch vinyl
The Twelve Hour Foundation *Quicksilver records* Castles in Space CiS144 2022, 7 inch lathe-cut vinyl
The Twelve Hour Foundation *The Hexagonal World of The Twelve Hour Foundation* Disques Boum BOUM19 2023, CD and 12-inch vinyl
there are no birds here *Ulfkell's Land* HREA'M Recordings HR Three, 2020, cassette (25 copies)
there are no birds here *Another Ruined Landscape* HREA'M Recordings HR Eight, 2021, cassette
there are no birds here *The Drift* Dub Cthonic 07, 2021 CDr

there are no birds here *Hume* not on label 2021, cassette (10 copies)
there are no birds here *Stryance* Courier Parcel1 2022, CDr
there are no birds here *riverrun,* HREA'M Recordings 2023, cassette (50 copies)
Thomas Ragsdale *Honley Civic Archives Volume One* Soundtracking the Void STV001 2018, download only
Thomas Ragsdale *Honley Civic Archives Volume Two* Soundtracking the Void STV005 2019, download only
Thorn Wych *The Legend of Willy Howe* (East Yorkshire) Folklore Tapes FTCCIV 2024, cassette issue of 100, b/w Adra *Ballad of the Linknsheere Commons* (Lincolnshire)
Tiny Leaves *Notes on Belonging* Pegdoll PEGX020 2017, CD
Tiny Leaves *Mynd* self-released 2023, download only
Town & Country *A Plan for Plymouth* Woodford Halse WF58 2022, limited edition cassette
Alexander Tucker and Keith Collins *Fifth Continent* Subtext SUB044 2022, 12-inch vinyl
Twilight Sequence *Outline of Nature,* Castles in Space CiS111 2023, 12-inch coloured blue vinyl
Twilight Sequence *Looking at Lifeforms,* Castles in Space SL-012 2022, 7-inch vinyl
Twilight Sequence *Trees in General: and the Larch* Castles in Space CiS089, 2022, 12-inch single-sided vinyl
Tyneham House *Tyneham House* Second Language SL015 2012, CDr with cassette, reissued Clay Pipe Music pipe_003 2016, 10-inch vinyl
Veil *A Circle in Stone* Other Facts 2023, booklet with download
Vic Mars *The Land and the Garden* Clay Pipe Music pipe_021 2015, 12-inch vinyl
Vic Mars *Inner Roads and Outer Paths* Clay Pipe Music pipe_021 2019, 12-inch vinyl
Vic Mars *The Beacons* Clay Pipe Music pipe_037 2023, 12-inch coloured vinyl, pipe_037CD, CD
Vic Mars *Bannau Brycheiniog* Clay Pipe Music pipe_041 2024, limited hand-made CD (100 copies)
Warmfield *Warmfield-cum-Heath* self-released 2024, cassette and zine
Warmfield *The Pinfold* Die Das Der records DDD59V 2024, limited lathe-cut 10-inch
Warmfield *Saver Strip* Die Das Der records DDD70CD 2025, CD
Warrington-Runcorn New Town Development Plan *Interim Report, March 1979* Castles in Space CiS067 2021, CD and 12-inch vinyl
Warrington-Runcorn New Town Development Plan *People & Industry* Castles in Space CiS088 2021, CD and 12-inch vinyl
Warrington-Runcorn New Town Development Plan *Districts Roads Open Space* Castles in Space CiS101 2022, CD and 12-inch vinyl
Warrington-Runcorn New Town Development Plan *Fig 1c The Third Phase of Housebuilding* Castles in Space CiS126 2022, 7-inch lathe-cut vinyl

Warrington-Runcorn New Town Development Plan *The Nation's Most Central Location* Castles in Space CiS127 2023, CD and 12-inch vinyl

Warrington-Runcorn New Town Development Plan *Building a New Town EP* Castles in Space CiS126 2023, 12-inch gold and autumn leaves coloured vinyl

Warrington-Runcorn New Town Development Plan *A Shared Sense of Purpose* Castles in Space CiS175 2024, 4-track 12-inch pink coloured vinyl and 2-track purple/cream 7-inch

Warrington-Runcorn New Town Development Plan *Community Hub* Castles in Space CiS157 2024, CD and 12-inch coloured vinyl

Wealdham *Motorway* Castles in Space CiS108 2023, limited edition 12-inch splatter vinyl

White Flowers *Day by Day* Tough Love Records TLV139LP, TLV139LPXX and TLV139CD 2019, 12-inch vinyl, limited ultra-clear 12-inch vinyl

Chris Whitehead *South Gare* Linear Obsessional LOR018 2012, download only

Wildly *Mynnd Tarw* Plenty Wenlock PW009 2023, CDr limited edition of 25

Mark Williamson, Mark *Stone Report Vol 1* Forged River FRCD011 2024, 20-page zine with 3-inch CD

Yggdrasil *9Worlds* Rusted Tone RTR0566 2022, cassette

Richard Youngs *Memory Ain't No Decay* Wayside and Woodland W&W044 2018, CDr

NB: unless otherwise indicated, all releases were made available for digital download in addition to the physical releases listed.

Index

www.ingramcontent.com/pod-product-compliance
Ingram Content Group UK Ltd.
Pitfield, Milton Keynes, MK11 3LW, UK
UKHW020216290926
15470UKWH00001B/211

9 798765 112922